The Real Estate Solar Investment Handbook

The Real Estate Solar Investment Handbook explains the business case for property professionals to pursue solar projects.

A project's value is determined by its potential risks and rewards; these are explained thoroughly in terms understood by the real estate industry.

This book provides a framework for practical decision-making, with each chapter addressing a step in the process, from project idea to completion.

Written from the perspective of the commercial real estate industry professional, it will help investors evaluate opportunities and execute projects that offer solid risk-adjusted investments.

For property owners, investors, landlords, service providers, and all those looking to invest in solar on commercial property, *The Real Estate Solar Investment Handbook* will guide you through all the steps needed to gain years of revenue from a project.

Aaron Binkley is Director of Sustainability Programs at Prologis, Inc. where he is focused on enhancing the energy performance of the company's industrial property portfolio. Prior to joining Prologis, he was Director of Sustainability Programs at AMB Property Corporation. He spearheaded the creation of a new rooftop solar energy standard that is slated for inclusion in the upcoming version of the US Green Building Council's LEED Rating System.

The Real Estate Solar Investment Handbook

A commercial property guide to managing risks and maximizing returns

Aaron Binkley

Routledge
Taylor & Francis Group

LONDON AND NEW YORK

from Routledge

First published 2014
by Routledge
2 Park Square, Milton Park, Abingdon, Oxon OX14 4RN

and by Routledge
711 Third Avenue, New York, NY 10017

Routledge is an imprint of the Taylor & Francis Group, an informa business

British Library Cataloguing in Publication Data
A catalogue record for this book is available from the British Library

Library of Congress Cataloging-in-Publication Data
Binkley, Aaron.
  The real estate solar investment handbook: a commercial property guide to
  managing risks and maximising returns/Aaron Binkley. – First Edition.
  pages cm
  Includes bibliographical references and index.
  1. Real estate business.   2. Solar energy.   3. Real estate investment.
  4. Risk management.   I. Title.
  HD1375.B556 2013
  333.792'3 – dc23
  2013015704

ISBN: 978-0-415-66038-9 (hbk)
ISBN: 978-0-203-36209-9 (ebk)

Typeset in Garamond 3 and Gill Sans
by Florence Production Ltd, Stoodleigh, Devon

MIX
Paper from
responsible sources
FSC
www.fsc.org   FSC® C013056

Printed and bound in Great Britain by
TJ International Ltd, Padstow, Cornwall

Contents

PART IV
Identifying and managing risks   125

PART V
Case studies   197

Figures

Tables

Preface

The "a ha!" moment for writing this handbook occurred during a phone call with a consultant that had worked with my company for many years. The consulting firm had been providing physical and environmental due diligence services for real estate acquisitions for many years and was highly regarded in the industry. The consultant explained that her team was performing due diligence for an acquisition opportunity, and for the first time had encountered a property with a large rooftop solar array. She proceeded to ask many basic questions about solar modules, system longevity, contracts, and rooftop impacts. As we spoke, it became clear that this highly experienced real estate industry professional had little knowledge of solar projects and their potential impacts on the underlying building.

What the consultant really wanted to know was: "What questions do I need to be asking so I can figure out whether solar is an asset or a liability for this property?" As this fundamental question sank in, it brought back other conversations I'd had with industry colleagues through the years, wondering if solar would work at their properties. They asked how to pick apart proposals they had received from solar developers, wondered how to respond to a tenant that asked to install a solar array, or inquired about how I had gotten management support for my solar projects. I realized that many of my real estate colleagues intuitively felt that there was value in solar projects – they just didn't know how to make sense of solar and put it into their real estate-focused frame of reference.

I wrote this handbook with the perspective of the consultant and my colleagues in mind. Its purpose is to provide the essential knowledge required to approach solar deals, evaluate them in conjunction with your professional experience in real estate, and make decisions about how to proceed. This handbook will arm you with the questions to ask and the critical risks to be on the lookout for. I expect you will bring a depth of real estate experience to the table, because in many ways solar projects can be viewed through a similar lens.

Every deal, every project in solar is different, as it is in real estate. There is no way to create a resource that addresses every possible project, property or market. By utilizing your own professional experience alongside the material in the handbook you will be able to confront the unique aspects of solar with confidence. The handbook is a resource to help answer the inquiry from our consultant – "What questions do I need to be asking?" Once you read this handbook and learn what questions to ask, your intuition and experience will enable you to dig deeper to get the answers that will enable you to make sound financial and operational decisions.

Acknowledgments

I owe an immense debt to Amber for all the airplane rides devoted to reading, commenting, and encouraging. I thank Andy for his steadfast and insightful counsel from project inception through to completion. I am grateful to both of you for your vital contributions to this work.

Abbreviations

AC	alternating current
BLS	Bureau of Labor Statistics
BREEAM	Building Research Establishment Environmental Assessment Method
BOS	balance of system
CASBEE	Comprehensive Assessment System for Built Environment Efficiency
CDP	Carbon Disclosure Project
CdTe	cadmium telluride
CIGS	copper, indium, gallium, and selenide
DC	direct current
DSIRE	Database of State Incentives for Renewable Energy
EPC	engineering, procurement, construction
EPDM	ethylene propylene diene monomer
EPIA	European Photovoltaic Industry Association
FiT	feed-in tariff
GRESB	Global Real Estate Sustainability Benchmark
IRR	internal rate of return
JV	joint venture
kW	kilowatt
kWh	kilowatt-hour
LEED	Leadership in Energy and Environmental Design
LLC	limited liability corporation
LP	limited partnership
MACRS	Modified Accelerated Cost-Recovery System
MW	megawatt
NOI	net operating income
O&M	operations and maintenance
PACE	property-assessed clean energy
PPA	power purchase agreement
PTC	PV-USA test conditions
PV	photovoltaic
PVC	polyvinyl chloride
REC	renewable energy certificate
REIT	real estate investment trust
SEIA	Solar Energy Industries Association
SEPA	Solar Electric Power Association

Si	silicon
sREC	solar renewable energy certificate
STC	standard test conditions
TPO	thermoplastic polyolefin

Introduction

This handbook provides a framework that helps you make informed decisions about how to pursue solar projects that maximize revenue and minimize risk. Because this book is written from and for the perspective of a commercial real estate professional, it honors the following assumption:

> If a solar project creates value to justify the added risks by generating revenue, lowering operating costs or attracting tenants, it's a good decision. Conversely, if the costs and benefits cannot be aligned, deploying solar may not be justified.

The world is increasingly recognizing the opportunity to generate clean, emission-free electricity by installing solar photovoltaic (PV) systems. Governments, local communities, and the business world are showing unprecedented levels of interest in using space on commercial buildings to host PV renewable energy systems. There are now opportunities for commercial property owners, developers, and operators to unlock new sources of revenue by hosting solar projects, often on under-utilized rooftop space. New solar project deal structures offer a path for property owners to generate attractive risk-adjusted returns while supporting the development of new sources of clean energy that reduce environmental impacts.

The potential for generating clean solar energy from rooftops is significant. By some estimates commercial building rooftops alone offer a resource of 30 billion square feet of space on which to install solar. If this space were to be fully built out it would have the potential to support as much as 300 gigawatts of solar capacity. When we look globally we see many countries with programs that support solar. Europe is currently the global solar leader, accounting for 70 percent of the worldwide solar market. The United States, China, and Japan make up the bulk of the remaining installed projects, although solar can be found in nearly every corner of the world.

Global companies that manage their own real estate portfolios have begun aggressively pursuing solar installations. Wal-Mart, IKEA, Toyota, Sainsbury's, Google, DHL, Apple, and FedEx are just a few of the many examples of global enterprises that have developed a strong business case to install solar projects. As a result they are deploying solar on commercial buildings at a never before seen scale. Sainsbury's, the United Kingdom-based grocery chain, has committed to installing solar on more than 115 stores. Wal-Mart had installed solar on more than 140 properties by the end of 2012. These projects are not just demonstration projects; each one has hundreds or thousands of solar modules and is intended to offset as much as 30 percent or more of each retail store's electricity needs. The world's leading companies have made the economic and business case to deploy solar projects – what have they figured out? Can those lessons also apply to the commercial real estate industry? If these companies happen to be tenants in your buildings, you may have already been confronted with this question.

This handbook focuses on providing the information necessary to allow commercial real estate professionals to understand how solar projects fit within the properties they develop, manage, and operate on a daily basis. While interest comes from many quarters within the real estate industry – I have heard similar questions from chief executive officers and property managers – they share a common goal of wanting to understand solar in order to be able to make an informed decision about how they can pursue it most effectively. The audience for this book includes:

- senior executives looking at solar as a strategic opportunity within their company;
- developers interested in adding solar to their development projects;
- fund managers deciding if solar can increase revenue for their fund;
- property managers seeking to understand the impact of a tenant's request for solar;
- buyers and sellers of properties seeking to accurately gauge the value of a solar project;
- the brokerage community marketing a property that has solar on the roof, and enhancing their ability to match solar companies with property owners;
- leasing agents charged with negotiating solar leases;
- real estate attorneys seeking to gain a better understanding of the solar contractual risks that they will have to resolve when representing their clients;
- professional service firms such as architects, engineers, and the construction industry seeking a better understanding of how to incorporate solar into their clients' projects;
- solar project developers looking to gain a better understanding of how they can serve the commercial real estate market;
- government officials looking to understand how to craft better policies to expand access to renewable solar energy within local communities;
- utility company professionals dealing with a growing number of commercial-scale solar projects in their service territory;
- solar product manufacturers seeking to develop new and better products that address the needs of commercial properties.

Written from the perspective of the commercial real estate industry, the handbook allows commercial property professionals to evaluate investment opportunities and execute projects that are solid risk-adjusted investments that extend well beyond opportunistic greenwashing. The lessons can be utilized for any asset type – office, retail, industrial, multi-family residential, hospitality, as well as many others outside of the commercial industry, such as municipal and educational sectors. When planned and executed correctly, solar projects can provide years of low-risk revenue for a property owner while producing a consistent source of clean energy.

The information in the handbook can be applied to a single property or to a global portfolio. At either of these scales, the opportunity for industry professionals will be based on many factors, including property square footage, asset type, and geographic location. These factors are influenced by the lease type, tenant mix, and operational characteristics of each property. Understanding how these factors affect your solar project can be a challenging task that competes for your scarce time and staffing resources. This handbook enables you to focus on the most relevant factors quickly and in a structured way in order to determine the right path forward. Once the most appropriate solution is determined, this handbook provides the information required to allow you to effectively manage risks when the project is being designed, installed, and operated.

This handbook is a user-friendly guide for real estate industry professionals. It assumes the reader is interested in learning about the solar industry in sufficient detail in order to evaluate solar opportunities and make sound business decisions. This book does not attempt to give you a science lesson on solar energy or electricity. It does not teach you how to build and install your own solar array. Instead, it focuses on the core knowledge you will need to:

- assess the solar opportunity;
- select the best contractual and financial project structure;
- identify and manage risks.

The handbook is divided into sections that correspond with these areas of core knowledge. The first two sections provide you with the tools to assess the solar opportunity and identify the value solar can bring to your organization. Whether you invest in solar systems is ultimately a business decision, but every commercial property professional should be aware of the economic, regulatory, and cultural trends that impact renewable energy today. A first step is to understand the most powerful drivers behind the value that results from deploying solar projects at your properties. The next topics the handbook discusses are the building blocks of solar projects; this provides an understanding of the key components of where solar will be a viable solution. This information also enables you to have an effective dialogue with service providers in the solar industry. With this knowledge at hand, the handbook describes a process that allows you to screen your property portfolio for solar opportunities. You will also learn how to identify solar industry resources, from technology options to solar service providers, that can provide support to help ensure the success of your projects.

The next section of the handbook identifies the major types of contractual and financial structures available to deploy solar projects. Contracts you negotiate must protect your property and manage risks effectively while achieving your revenue and value creation objectives. The information contained in this section allows you to identify the most compatible project structure for your business and solar goals. Each chapter arms you with the key deal points for each contractual structure that enables you to negotiate an agreement that protects your interests and ensures a successful outcome.

The last part of the handbook explains how to identify and manage risks that you may encounter during the construction and operation of your solar project. Working through the due diligence, construction, and operational requirements for your solar project is one of the most important steps to ensuring long-term value. The chapters in this portion of the handbook identify specific risk factors and explain solutions to allow you to manage them effectively. In this section you will learn how to identify physical and operational risks, and how to manage them. You will also be introduced to industry-wide systemic risks that are outside of the direct control of the project. Managing these risks effectively is essential for delivering successful solar projects.

The handbook walks you through a process that gives you an understanding of how a project takes shape. Each section describes the key factors that affect the long-term value of solar projects. The organization of the handbook is intended to make it efficient to identify what opportunities exist and how to capitalize on them while managing risks effectively. This is the secret to creating lasting economic value with solar projects.

Part I

Assessing the solar opportunity

When you begin to look at the opportunities solar offers for your property portfolio, there is often a moment when the stream of solar company solicitations and news becomes overwhelming. In one moment you hear that solar is the way of the future; later on the same day you read that solar is a costly vision that has little place in the real world. I am not surprised when I talk to industry colleagues and hear comments like this:

- "I want to be clear on what it takes to deploy a quality solar project."
- "I want to understand the difference between what solar companies are offering."
- "I want to know how to quantify the value solar creates for my business or property type."
- "I want to know where to find resources and support."

This is what often happens: A commercial property professional decides to take a look at one or two properties with a single solar company that proposes one type of solar contract. If an agreement doesn't come together at that moment in time, the decision is made that solar is just not a good fit for their portfolio or business.

But these same people would not decide to enter a new real estate market, attempt unsuccessfully to buy one building, and then give up and leave the market. They would have a long-term strategy and a systematic approach to build an understanding of the market, target properties, and position themselves to capture the greatest value over time. The pursuit of value from solar projects requires much the same approach. Market knowledge, clarity around your goals, and the ability to capitalize on changes in the market over time can help effectively mitigate solar risks while maximizing value.

With this in mind, Part I of the handbook introduces the knowledge that is essential to understanding how solar projects create value for commercial properties. This information allows you to identify the components of value that solar projects create that are most relevant to your goals and operational needs.

You will learn the four key components that must be in place from a financial and functional standpoint for solar projects to be viable. Building on the four components, you will learn how to conduct due diligence within your real estate portfolio to identify the properties that will be the best candidates for solar projects. This will enable you to target the markets and properties that will be the best candidates for solar and that create the most value for you.

Chapter 1

Key terms used by the solar industry

Chapter summary

- Understanding solar terms and phrases makes your conversations with solar companies and service providers more productive and reduces the potential for confusion.
- Knowing the language of the solar industry makes it easier to analyze your options and determine where the most value exists.

Each industry has its own language, and the solar industry is no different. The solar industry has numerous terms that allow you to understand information and communicate effectively with solar service providers. Many of these terms may already be familiar to you. Following is a highly abridged summary of key words and phrases that explain the most relevant terms you are likely to encounter. Rather than being overly technical, the terms are described in a practical and comparative way so you can understand what the equipment is and understand it in context.

Alternating current, abbreviated as "AC," is the form of electricity used by buildings and the utility grid. The direct current output of solar arrays is converted to AC before being delivered to the building or utility grid.

Balance of System, abbreviated as "BOS," is the materials and equipment required to make a complete and functional PV facility, excluding the solar modules. This includes the racking system, ballast, wiring and conduit, conductors, electrical equipment, and the like. The term is commonly used in context as "balance of system components."

A *building-applied* solar array is a PV system affixed to the building, such as the roof surface or a façade. A building-applied solar array can be removed without dismantling portions of the building.

A *building-integrated* solar array is a PV system permanently constructed as part of a standard building component such as curtain wall glazing, exterior cladding or roof membrane. A building-integrated solar array cannot be removed without dismantling portions of the building.

CdTe is the acronym for *cadmium telluride*, the elements that comprise the semiconductor used in certain types of thin-film PV cells.

CIGS (pronounced "siggs") is the acronym for *copper, indium, gallium, and selenide*, the elements that comprise the semiconductor used in certain types of thin-film PV cells.

A *curtain wall* is a type of cladding system used widely for commercial buildings. A curtain wall consists of an aluminum or steel frame that is hung from the primary structure of the building. Curtain walls function as the exterior enclosure of the building and often contain glass within their frames that acts as windows. Metal, stone, or panels of other materials may also be used in a curtain wall frame. Solar modules can be designed to fit into a curtain wall frame.

Degradation of solar modules is the amount of performance lost in each year of operation. High-quality solar modules are generally warranted to retain 80 percent of their rated output after 25 years.

Direct current, abbreviated as "DC," is the form of electricity produced by PV modules. Direct current electricity generally must be converted to alternating current before a building can use it, or before it can be sent back to the utility grid.

Disconnects are switches, also referred to as circuit breakers, that cut off power flowing through them in the event of a problem with the solar array. There is typically a disconnect on either side of the inverter. This allows the solar array to be isolated from the rest of the system for repair or diagnostics. Disconnects are also used to isolate the solar array from the building or utility grid if the electrical infrastructure of either one needs to be repaired.

An *electrical enclosure* is a dedicated space, usually outside on the ground or inside a building electrical room where the electricity-handling equipment for a solar facility is placed. This is where the inverter, transformer, disconnects, and metering equipment are typically located.

A *feed-in tariff* (FiT) is a term used by the energy industry to describe a fixed payment contract, typically between a power producer and a utility. The contract stipulates that the purchaser will pay the energy producer for each unit of energy that is generated and delivered to the grid. A FiT can be used for numerous sources of energy such as wind or biomass, not only solar. The price of a FiT varies based on location and regulatory policy, but FiTs are generally based on the cost of generation of each technology. FiT pricing is periodically adjusted by regulators to reflect changes in the cost of developing projects and generating electricity.

Flexible thin-film modules have an appearance that differs from what most people associate with traditional solar panels. They are a thin, flexible module about 16 inches wide and up to 20 feet long. The PV cells are laminated between sheets of durable plastic laminate atop a thin stainless steel backing. This type of module does not have any frame and does not require a separate support structure. It can be rolled out and affixed directly onto many different types of roof surfaces. This module generally has lower efficiency than traditional crystalline solar modules, but it is lightweight.

An *inverter* and transformer, along with other electrical components, turn raw electricity from solar modules into a form usable by the building or the utility grid. The inverter converts electricity from DC to AC. The electricity is then run through a transformer that adjusts the voltage to make it compatible with the building or the utility grid. Inverters come in several size categories, as shown in Table 1.1.

Irradiance is the amount of solar radiation striking the earth's surface, measured in watts per square meter (W/m^2).

Irradiation is the amount of solar radiation striking the earth's surface for a given amount of time, measured in watt-hours per square meter ($W\text{-}H/m^2$).

Kilowatts, abbreviated as "kW," are a measure of electric power at any given instant. For example, a solar array could produce anywhere from zero kW at night time, to many thousands of kW at midday; 1 kW = 1,000 watts.

Table 1.1 Inverter comparison

Type	Size	Location
Micro inverter	One per solar module	Attached to each solar module
String inverter	One per array (typically 50 kW or less each)	Attached in rows to a wall or other supporting surface
Central station inverter	One per project (typically from 100 kW to more than 1,000 kW)	Free-standing in an electrical enclosure

Kilowatt-hours, abbreviated as "kWh," are a measure of energy, calculated as power × time. For example, 1,000 kilowatts consumed during the course of two hours equals 2,000 kWh.

A *megawatt*, abbreviated as "MW," is a measure of electrical power; 1 MW = 1,000,000 watts. The capacity of large solar arrays is often described in MW, such as "a 2 MW rooftop solar project."

A solar *module* is an assembly of PV cells and a supporting structure that form a panel. Solar modules are commonly referred to as "solar panels." Modules are usually rectangular, measuring about 2' × 6' on a side. These modules contain rows of PV cells sandwiched between sheets of plastic laminate and glass. A module has electrical connectors built into the side facing away from the sun to interconnect with other modules to form an array. Some modules have an aluminum frame around the perimeter of the module, while others do not.

Mono-crystalline solar cells describe a type of silicon semiconductor used to make solar modules. These types of cells tend to have a high efficiency. Mono-crystalline solar cells are dark blue or black in color and are made from highly refined silicon, an abundant natural material found in sand and quartz.

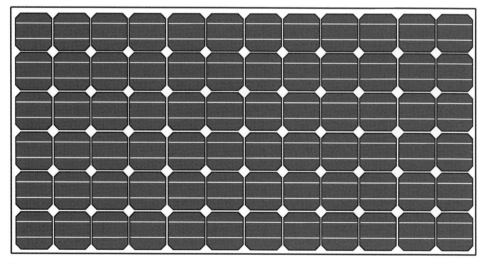

Figure 1.1 Solar module.

Note: This module is comprised of 72 photovoltaic cells.

The *nameplate rating* is the rated capacity of a solar array, calculated by summing the wattage of all solar modules in the array, typically expressed in kW. A solar array composed of 100 modules rated at 250 watts has a nameplate rating of 25 kW. See also *rated system capacity*:

100 modules × 250 watts = 25,000 watts, or 25 kW

Net metering is a regulatory policy that exists in most property markets. This policy is what enables property owners to connect their solar system to the utility grid. The term refers to the "net" amount of electricity that the utility charges each month once solar production is subtracted from the total metered consumption of energy from the grid.

Photovoltaic cells, abbreviated as "PV" and often referred to as solar cells, are the individual components in a solar module that generate electricity. Cells measure roughly six inches on a side, and are composed of semiconducting materials that generate DC electricity when struck by the sun. PV cells may be composed of silicon, as well as combinations of elements including CdTe, CIGS, as well as other compounds. PV cells are combined in a sheet to form a module.

Polycrystalline solar cells are a type of silicon semiconductor used to make solar modules. These types of cells tend to have a moderately high efficiency.

A *power purchase agreement*, abbreviated as "PPA," is a long-term sales agreement between a producer of energy and a buyer of energy, where the producer agrees to sell the output of their solar array to the buyer at an agreed-upon price for a fixed period of time.

Racking is a support structure for the solar modules in an array. Racking holds the modules in a predetermined position facing the sun, while anchoring them to the building. This can be a permanently anchored system or a ballasted attachment, and it prevents the modules from moving out of position. Racking is usually fabricated from stainless steel, aluminum, or high-strength molded plastics. Some specialized systems are integrated into building façade elements, such as the curtain wall. In this

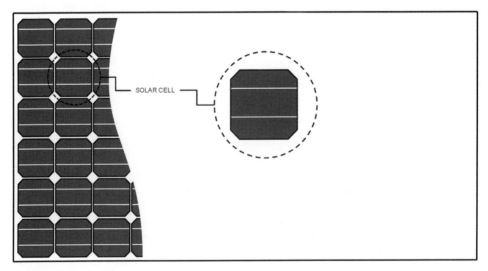

Figure 1.2 Photovoltaic cell in a solar module (cutaway).

case the solar modules are used in place of the glass in between the metal framing members. Racking systems may be assembled from parts in the field. This is referred to as "field assembled" or "stick-built." This allows the racking to be customized for unique conditions found at the job site. Sometimes a portion of the racking system will be prefabricated off-site to reduce installation time.

Rated system capacity describes the size of a solar array. This is a number in kW or, for very large projects, in MW. This number is determined by adding up the rated capacity of all the solar modules that make up the project. See also *nameplate rating.*

Renewable energy certificates, abbreviated as "RECs" and pronounced as "wrecks," represent the intangible emission-free attributes of the electricity produced by a clean energy source. A solar renewable energy certificate is technically described as an "sREC." The standard unit for a REC is 1 MW-hour of energy, produced by a renewable energy facility such as a solar array or a wind energy farm. RECs can be traded in a renewable energy certificate market in certain regions. The energy produced by a solar array and the renewable energy attributes conferred by the REC can be disaggregated; that is, sold separately. The value of RECs can vary widely from one market to another. In certain markets, RECs have virtually no value, while in others they can be more valuable than the electricity produced by the solar array.

Soiling is a term used to describe the process of dirt and other foreign matter collecting on the surface of solar modules after they are installed. As soiling increases, electricity output declines because less sunlight can reach the solar cells. Soiling is caused by dust, dirt, bird guano, as well as other wind-blown matter that sticks to the surface of the solar modules.

A *solar array* is a series of solar modules connected to one another. The term "array" is used in the solar industry in a few ways. It can describe all the modules in the system, or it can refer to a subset of modules that are grouped together within a larger project. For example, a 30 kW solar project may be referred to as a "30 kW array." Large projects typically have many arrays that are building blocks of the entire system. A 500 kW rooftop system may be composed of ten interconnected arrays, each 50 kW in size.

Standard test conditions (STC) refers to the official standard for measuring rated capacity for an individual solar module. This allows you to compare the rated power output of one module to another. These ratings are, with a few exceptions, shown in increments of 5 watts – i.e. 200 watts; 205 watts; 210 watts. Solar module product specification will often list STC as "power output," "nominal power," or "Pmax."

STC is useful for comparing rated power of modules, but it doesn't describe how the modules will perform in the real world. For this, the solar industry uses an acronym referred to as "PTC." PTC is an abbreviation for PV-USA test conditions. This measure was developed to approximate the performance of modules under real-world conditions. PTC is not always listed on module spec sheets but the manufacturer can provide it. PTC is generally about 10 percent lower than STC, although the difference can be greater than that for certain modules. You can find a comparison of STC to PTC values for an extensive inventory of solar modules on the Go Solar California website at: www.gosolarcalifornia.ca.gov/equipment/pv_modules.php

System output is measured in kWh. One kWh equals 1,000 watts of electricity produced for a period of one hour. In this example, one kWh would be sufficient to power ten 100-watt light bulbs for one hour. Solar arrays with the same rated capacity can produce different amounts of kWh. This variation is based on location, system efficiency, and weather conditions where the array is installed. For example, a 30 kW rated solar array using 18 percent efficient solar modules will produce more electricity than a system of the same rated capacity that uses 14 percent efficient modules. Similarly, a 30 kW array in a sunny region in Mexico is likely to produce more electricity, measured in kWh, than an identical array installed in Canada, where the sun's energy is less intense.

A *thin-film* solar module is a general designation for PV cells that do not use crystalline silicon. Examples include amorphous silicon and CdTe solar cells. Thin-film modules may or may not be flexible. Practically speaking, though, there is little difference in the thickness of fully assembled solar modules regardless of whether they are thin film or not. All PV cells need to be encapsulated in glass or plastic in order to be made weathertight and compatible with racking systems.

Tilt is a term that describes the angle at which solar modules are installed relative to a horizontal surface. Modules are tilted to face the sun more directly in order to increase their performance. A tilt angle of 30° means the solar module is tilted 30° from the horizontal surface. Solar modules may be installed at any angle. Many racking systems are pre-configured to hold modules at one or two commonly used angles, such as 5° and 10°, or 20° and 30°. Some modules for rooftop projects are manufactured and shipped to the site with an integrated thermoplastic racking system that positions the modules at a fixed 5° tilt.

A *transformer* is an electrical device that converts incoming electricity from one voltage to another voltage. Solar arrays may have a transformer to convert the electricity produced by the solar modules to either the same voltage as the underlying building or the utility grid.

Walk pads are thin sheets or mats of material installed on the surface of a building roof to protect the roof surface from foot traffic. Walk pads are commonly used in areas where foot traffic is concentrated, such as at roof ladders, around mechanical units, and in pathways between rows of solar arrays.

A *watt* is a measure of power. Solar modules have a power output capacity described in watts, such as "a 240-watt solar module."

Wind uplift is the force generated by air moving across a solar array, which creates areas of low pressure that tend to "lift" the solar modules off of the building. The racking system, including ballast and anchors, is designed for each project to resist these forces. Wind uplift is generally greater in higher wind-speed zones in coastal and other windy regions. Wind uplift forces also tend to rise as the tilt angle of the solar modules increases.

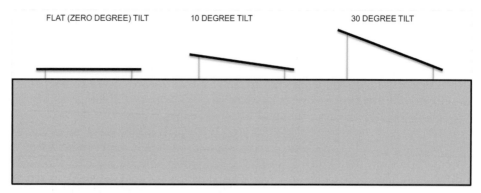

Figure 1.3 Tilt of solar modules.

Chapter 2

Six elements of solar project value

Chapter summary

- Solar industry growth has spurred new ways for projects to generate revenue for commercial properties.
- Solar offers a revenue arbitrage, allowing property owners to generate income from a previously unvalued asset: the roof, walls and parking spaces of a building.
- Adding solar to a commercial building reduces energy expenditures and can lock in stable electricity costs.
- Incorporating solar into development plans enhances regulatory and community support for projects by demonstrating a commitment to reducing environmental impacts.
- Adding solar to existing buildings appeals to investors that have socially responsible investing policies.
- Delivering solar energy to your building can attract and retain customers seeking sustainable workplaces and a smaller environmental footprint.

There are six ways in which solar projects create value for commercial properties. These are predominantly focused on financial value; some of them also provide value in non-monetary ways. These elements help you outline a business case for pursuing solar projects, where the value created by solar justifies the risks you take on. It is important to recognize that what has value for one property owner may be different for another. For some, the value may be purely based on generating revenue from under-utilized rooftop space. For others, using solar to reduce costly peak energy demand charges may create significant value. Still others may find that adding solar during the planning stages of a development project lowers their carrying costs by strengthening community support and speeding up the permit process. For most property owners solar has the opportunity to create value in more than one area.

Industry drivers

It is important to understand the evolution of the solar industry over the past few decades. Two, or even one, decades ago in most markets there were not six elements of solar project value, there were only two or three. This rapid evolution of the solar industry has created opportunities for solar projects to add value to commercial buildings that did not exist in the recent past. This positions solar as an opportunity for more and more commercial buildings. Before identifying the six elements of solar project value, it is helpful to understand what has changed in the solar industry to bring solar to the table as a source of value for commercial property owners. These industry changes include the following.

Enhanced regulations

Regulations in many countries have enabled solar projects at a scale never before seen. Until the past decade or so, regulations in many markets actually prohibited the installation of commercial-scale solar installations. This has changed rapidly as utility rules and government incentives have evolved to support renewable energy. Today some of the greatest industry growth is in large commercial-scale and utility-scale projects installed on buildings and on the ground.

Businesses and property owners now have an incentive to install solar projects that do more than just look good and provide a marketing opportunity. Today's projects generate revenue, reduce operating costs and increase property valuation. The economic incentive to pursue larger systems that meaningfully benefit commercial buildings and their owners now exists in many parts of the world.

Government support

In the past decade, many national governments have greatly expanded long-term support for solar energy projects. This is an outgrowth of consumer demand for cleaner sources of energy, a desire to lessen dependence on fossil fuels, and concerns about climate change. When Germany enacted a feed-in tariff (FiT) for solar power and streamlined utility rules, it kicked off a decade-long growth spurt that has transformed Germany into the world's largest solar market. Many individual countries and states have followed suit and created solar incentive programs for businesses and property owners that install solar projects.

California has become the largest solar market in the United States by reforming utility rules and enacting aggressive renewable portfolio standards that foster utility company programs that support solar systems. In the past few years even the armed forces have become strong advocates for renewable energy and major purchasers of projects. In 2012, the United States Army committed to a goal of producing one gigawatt of renewable energy from solar and other sources – enough energy by military calculations to power more than 250,000 homes.[1]

Quick tip

A FiT is a solar incentive payment where a solar project owner is paid a set rate for each kilowatt-hour (kWh) of electricity delivered to the utility grid.

Technology

Technological advancements have made solar panels significantly more efficient, while also reducing their cost by more than 80 percent in the past decade. Exponential growth in industry capacity combined with globalization of manufacturing has expedited movement down the cost curve. Improved solar array designs mean more sites can accommodate even more modules than before. Rising efficiencies of solar modules means that projects produce more energy at lower and lower costs. The growth in the number of experienced installers has further reduced costs and increased the quality of projects that get built.

Financial innovation

Financial innovation has fostered the growth of the solar industry. The financial industry is evolving to support solar projects at a scale not previously seen. Both debt and equity are now available, targeting high-quality projects. Financing solutions such as power purchase agreements have been reinvented for the solar industry and now make up the predominant product offering for many solar companies. Major financial institutions such as Goldman Sachs, JP Morgan, and Deutsche Bank have introduced solar investment funds and solar equipment leasing programs.

Further innovation is occurring with solutions such as property-assessed clean energy (PACE) financing for solar and energy-efficiency projects. This solution enables the financing for a solar project to be carried on the property tax bill – enabling a solar project to be passed from one owner to the next over the life of the solar project. Solar renewable energy certificate (sREC) markets also now exist, where the value of emission-free energy is traded on a daily basis. These innovations have made capital available for solar projects at a scale never before seen by the commercial real estate industry.

Figure 2.1 A large rooftop solar project in California.

Quick tip

A renewable energy certificate, or REC (pronounced "wreck"), represents the clean attributes of one megawatt-hour (1,000 kWh) of solar electricity. RECs vary in value from virtually zero, where there is no market to trade them, to being a primary form of incentive value in markets where their purchase is mandated by utility regulators.

Energy prices

Electricity prices continue to rise throughout the world. This helps to make commercial-scale solar systems viable in two ways. First, solar arrays can be used to offset increasingly costly grid-supplied power, particularly on days when peak summer time-of-use charges make buying electricity from the grid extremely expensive. Second, selling solar energy from large commercial-scale projects on a wholesale basis to utilities in high-cost energy regions is becoming economically feasible in an increasing number of markets. As more utilities transition to smart metering and time-of-use energy pricing models, the marginal value of solar energy production is poised to increase even further.

Six elements of solar project value

With these changes in the global solar industry as a backdrop, this section identifies the six most powerful sources of value for the commercial real estate industry. These include both direct financial benefits as well as intangible reasons driving commercial property owners' growing interest in solar projects. As you read through this section, consider which of these could create value for your business and be compatible with your operational needs.

1 Revenue arbitrage

Property owners have the ability to install solar facilities on buildings that were not originally developed or acquired with solar in mind. Solar projects can take advantage of financing solutions available in the solar industry to generate revenue for the property without an offsetting capital investment to purchase solar equipment. This creates an arbitrage opportunity – to create a new source of revenue without making a financial investment. The owner of a warehouse in southern California increased annual net operating income by more than $150,000 with no out-of-pocket capital investment by leasing the rooftop space to host a solar project. This project has added more than two million dollars in value to a building that was not built with solar in mind. This arbitrage opportunity will continue until developers and property owners include solar project investments in their underwriting.

2 Energy price reduction

Population growth and utility deregulation of many markets has exposed property owners to electricity rate increases and rising demand charges for energy consumed at

Figure 2.2 A mid-size solar array.

peak periods of the day. Increasing electrical infrastructure costs to replace outdated systems and power plants further contributes to rising prices. Commercial properties in many markets are being placed on electricity rate schedules that include time-of-use pricing where energy used at peak periods is much more expensive than energy at off-peak times.

Solar offers a means for property owners to better manage energy costs by locking in a fixed price for solar electricity from on-site solar projects. On-site electricity production offsets a portion of the energy that would normally have been purchased from the grid. Just as importantly, peak solar system output tends to coincide with times of the day and year when time-of-use charges are highest. Structuring a solar agreement where the cost of the energy it supplies is fixed at a price equal to or lower than today's electricity rates offers a hedge against rising energy prices and time-of-use charges. For example, a large cold-storage facility in New Jersey installed a 9 MW rooftop solar array to offset energy use at its chilling facility. By locking in a fixed price for electricity produced by the rooftop solar array and significantly reducing peak demand charges, the facility reduced its utility costs by more than one million dollars per year.

3 Energy and sustainability regulations

Municipalities responsible for permitting real estate developments are placing an increasing burden on property owners and developers to manage the impacts – real or anticipated – on the local environment. This has been the case for years with other environmental impacts such as wetlands restoration, open space preservation, and endangered species habitat protection. In recent years many developers have seen requirements for green buildings that tread more lightly on the environment.

In an increasing number of markets, developers and property owners are beginning to face regulations that mandate the integration of renewable energy into new commercial buildings. London, England has regulations in place today that require new commercial buildings to generate energy on-site from renewable sources. California is exploring regulations that would require all new commercial buildings to be zero-net energy by the year 2030. Zero-net means that new commercial buildings will be designed to produce as much clean energy as they consume over the course of a year. Similar regulations are being explored in New York and in countries around the world.

Sustainability-focused property developers are increasingly refining their specifications so that new projects incorporate renewable energy. This not only attracts build-to-suit customers that have a strong sustainability commitment, it also enhances regulatory and community support for the development. This can decrease the time required to complete a development, reduce carrying costs and allow capital to be recycled into future opportunities more quickly. For example, in the city of Chicago, Illinois projects that meet pre-established sustainability criteria automatically receive expedited permit review, which can reduce project schedules.

Whether mandated by building codes or implicitly required in order to obtain timely permit approvals, green buildings are being used by municipalities to balance development needs with environmental concerns. Incorporating solar into broader sustainability measures provides a unique solution for developers to meet a municipality's environmental objectives while also enhancing the property by generating revenue and reducing operating costs. These situations foreshadow a future for real estate development where renewable energy knowledge becomes mandatory in order to deliver projects that local communities want.

4 Responsible property investment

Investors that fund real estate development projects and that invest in stabilized real estate assets increasingly want to know that the properties they are investing in align with their socially and environmentally responsible investment policies. This is becoming prevalent among European investors when deploying capital, both inside and outside of Europe. They view sustainability integration as a fiduciary responsibility. The Global Real Estate Sustainability Benchmark (GRESB) is emerging as a platform for investors to assess the environmental sustainability performance of the funds they invest in. More than 450 global real estate companies and investment funds representing $1.3 trillion in assets under management have responded to GRESB at the request of institutional investors.[2] This survey's scoring system factors in a broad range of sustainability metrics, including the amount of solar on buildings in the portfolio and the growth, year-over-year, of the solar portfolio.

Another driver of responsible property investment is the Carbon Disclosure Project (CDP). The CDP is a leading advocacy group for responsible investment that has support from more than 722 institutional investors with combined assets under management of more than $87 trillion.[3] The responses to CDP include disclosure of the number of solar energy installations and renewable energy production.

Figure 2.3 A free-standing carport canopy solar array.

Institutional investors are taking a closer look at the sustainability investments in their portfolios, and that includes solar. Socially responsible investors are using the level of sustainability integration as a proxy for fund management quality to help tip the balance in deciding where to allocate future investments. Addressing investor interest proactively can make it easier to attract and retain investment capital. Solar energy projects are a valuable component of a strong sustainability program that demonstrates responsiveness to the sustainable objectives of investors.

5 Occupant demand

Building occupants increasingly seek ways to reduce their operating costs. Solar projects can offset energy costs and reduce the variability of future energy prices by locking in a fixed price for solar-generated electricity. Procuring solar energy is a way these companies can demonstrate that they are focused on sustainable practices that strengthen their brand. For the property owner, being able to offer occupants a lower energy cost supports lower operating expenses compared to other buildings. This helps to attract and retain tenants. This is particularly true where the tenant signs a long-term contract to purchase the energy from a solar project or makes an investment in their own solar facility at the property.

Even in situations where solar is only powering common area energy needs and not the tenant's own energy use, the image of the property owner and the building may be enhanced in the eyes of the tenant.

As an example, the news service Bloomberg occupies a historic waterfront office building in San Francisco. In 2012 they commissioned a rooftop solar array that was installed in collaboration with the property's landlord. The solar project enhances the existing LEED-certified office space for Bloomberg, and it provided an opportunity for the landlord to nurture their relationship with a valuable tenant. LEED, which stands for Leadership in Energy and Environmental Design, in a US-based and widely used sustainable building rating system.

6 Reputation

Solar projects can improve a commercial property owner's image as a responsible member of the business community, strengthen their perception among customers, and enhance a reputation as an employer-of-choice within the real estate industry. These branding opportunities are a way to differentiate properties, management, and services for customers. You are also likely to find that your own employees appreciate the efforts their company is making to reduce environmental impacts. This has proven to be a surprisingly strong factor in employee perception of its employer. In one large real estate company, an employee survey found that sustainability was a leading factor in employee satisfaction for 80 percent of employees. I have heard from numerous employees how gratifying it is to be able to go home and tell their family about the ways their company is doing good things for the environment. From the company's perspective, being seen as an employer-of-choice can create significant value by simply reducing costly employee turnover.

Some companies have used renewable energy to burnish their public image. For example, Google is one of the world's largest data center users. Data centers, even energy-efficient ones, consume vast amounts of electricity. One of the ways Google has deflected criticism for their huge energy footprint has been by installing large solar installations at several data center sites. They are also investing hundreds of millions of dollars in large-scale renewable solar and wind energy. This approach has value on several levels, but there has been a clear decision on their part to use solar to soften the image that their data centers are large energy consumers. And it is working. Google has since been recognized as a leader in the Greenpeace Cool IT Leaderboard for their commitment to supporting renewable energy such as solar and wind.[4]

Final thoughts

Of the six major drivers of value from solar projects, some may have been easy to guess, such as generating revenue from selling electricity or using the branding opportunity of solar to strengthen corporate reputations. Others, such as reducing carrying costs for development projects, are less obvious but no less important to consider when determining the ways solar projects can create value. And there are likely to be even more opportunities to derive value from solar in scenarios that are particular to your properties and your business. New drivers of value not listed in this chapter are also likely to emerge as the industry continues to evolve. Regardless of what property type you are focused on or what objectives you have for pursuing solar, there is almost certain to be one or more ways that solar can add value to your business.

Notes

1   "Army commits to security through renewable energy," www.army.mil/article/80303/Army_commits_to_security_through_renewable_energy
2   As of October 17, 2012: www.gresb.com
3   As of March 25, 2013: www.cdproject.net
4   www.greenpeace.org/international/en/publications/reports/Cool-IT-Leaderboard-5

Chapter 3

The building blocks of solar projects

Chapter summary

- Successful solar projects require a combination of four factors: insolation, electricity price, solar technology, and incentives.
- Insolation, the amount of the sun's energy at a given location, affects the amount of electricity that can be generated by a solar array.
- The price of electricity from the utility grid affects the value of solar energy; as prices rise the value of solar increases as a substitute for energy from the grid.
- The choice of solar technology affects the cost and energy output of a solar array.
- Solar incentive programs are designed to spur demand for solar projects, and are administered through government programs and utilities. These incentives include cash grants, tax credits, production-based payments such as feed-in tariffs (FiTs), and non-monetary project support such as expedited permitting.
- Incentive programs can be identified by working with solar companies, real estate service providers, utilities, and solar industry consultants.

This chapter provides the recipe necessary to create economically viable solar projects. This recipe has four basic ingredients:

1 insolation;
2 electricity price;
3 technology;
4 incentives.

Not long ago I took an informal poll among my colleagues, asking them to name the factors most important to making a solar project viable. Nearly everyone said that having abundant sunshine was key, and many also identified high electricity prices as a factor in making solar successful. Only a few mentioned the choice of solar technology or financial incentives. While you can – and will – have the four ingredients in different quantities depending on where your project is located and how it is financed, it is very difficult to make a solar project work if you are entirely lacking one or more ingredients.

As my informal poll found, abundant sunshine and high electricity prices may appear to be the most essential, but they are actually only a portion of what is needed for a successful solar project. In fact, in many cases today these are less important than having an appropriate level of incentives. Compared to the other factors, incentive levels have the most flexibility to be adjusted to balance out the other ingredients.

For example, solar project incentives in France were lowest in sunny southern France and progressively increased as they headed north, where there was less sunshine, in order to support projects throughout the country. Similarly, incentives on the islands of Hawaii have tended to be low because the cost of electricity is high – more than twice the national average in the United States. In this market, electricity prices play a leading role in making solar projects feasible. Looking at the markets that have had consistent success achieving solar market growth, you will find that they have created a recipe that balances these ingredients in ways that reflect the needs of their market.

An incorrect mix of ingredients can in some cases quickly overheat, or in other cases stifle, a solar market. Both scenarios are counterproductive to the goal of developing a stable solar market where property owners have certainty that their projects can be executed without undue risk of sudden market changes. We see this in the example of the Spanish solar market. In 2007 Spain introduced a lucrative FiT incentive. The market saw project applications grow out of control in 2008 as solar developers sought to cash in on the generous incentive program. The following year, incentives were trimmed back significantly. This was done in order to cool the market and to limit the government's ballooning incentive payment obligations. From 2008 to 2009, the number of megawatts installed dropped almost 80 percent, from 2,400 MW to 500 MW. In this case, Spain had abundant insolation, access to efficient and reasonably priced solar technologies, and moderate electricity prices. Adding high incentive levels was unnecessary and quickly proved to be unsustainable. The ensuing incentive cuts were severe and the market has been heavily constrained ever since.

Insolation

Insolation is the combination of intensity and duration of energy from the sun in a given location. To oversimplify a bit, you can think of insolation as the amount of sunshine in a particular location. More insolation means more energy is delivered to the solar modules. This, in turn, affects how much electricity the solar modules can produce. A solar module installed in a sunny climate such as southern Italy will produce more electricity than the same module installed in a region that is more northerly and cloudy, such as the United Kingdom. Insolation in the sunny southwestern United States can be 30–40 percent greater than in overcast northwestern states. This is one of the reasons for the rapid growth of utility-scale solar projects in the American southwest.

You can look up the insolation levels for the locations you are considering for solar by using the data and maps on the following solar resource websites:

- United States: www.nrel.gov/gis/solar.html
- Europe, Africa, and Asia: http://solargis.info/doc/71

Insolation levels increase the closer you get to the equator because the sun's energy is delivered most directly there. Projects at higher elevations also receive more energy from the sun. All things equal, a solar module in Mexico City, at an elevation of 8,000 feet (2.4 km) above sea level, will produce more energy than the same panel a few hundred miles due west at sea level along the Mexican coast.

If the properties you are considering for your solar project are in a not-so-sunny region, don't despair. As previously noted, insolation's impact on a project's economic viability can be balanced out by modest changes in incentive levels. This is well known and is factored into the design of incentive programs. For example, when France launched their solar incentive program they created levels of incentives to account for variations in insolation levels throughout the country. Incentives are lowest in the sunny Mediterranean climate in the south of France, and are highest in the overcast northern Atlantic regions.

Quick tip

Insolation is a measure of the intensity and duration of the sun's energy in a given location. Insolation tends to be higher in sunny climates closer to the equator and lower in more northern geographies.

Electricity price

The price of grid-supplied electricity affects the cost-effectiveness of solar projects. Electricity produced by a solar facility may be used to reduce the quantity of electricity purchased from the grid, or it can be sold back to the utility. Which of these options is available depends on the regulations in a given market. Solar energy can be used to offset building energy needs directly in many markets.

Where electricity costs are high, the savings generated by a solar array can readily be calculated by determining the avoided cost of purchasing energy from the grid. Where energy can be sold to the grid, high electricity prices signal that the energy from a solar array is likely to have a higher corresponding value than in a market with lower energy prices. This relationship is determined by regulations and policy, however; calculating this precisely will vary from one market to the next.

Solar power is particularly valuable in locations where electricity prices are high, and for buildings that have high demand for energy at peak periods during the day when electricity is most expensive. Commercial buildings often pay for both consumption measured in kWh, and also for time-of-use measured in kW to account for instantaneous usage. This peak kW time-of-use charge is more costly than baseline kWh consumption. In markets where energy resources are constrained – many major urban areas – these demand charges can be many times higher than the baseline kWh charge. Not surprisingly, demand charges are highest when the need for electricity is greatest on hot summer days when cooling systems are trying to keep up with high temperatures. These are the times when solar arrays are at their best.

This creates a higher value for every unit of electricity produced by the solar array because it not only offsets the amount of electricity consumed, it also reduces the level of peak demand that would otherwise be needed from the grid. In locations where tiered time-of-use charges become increasingly expensive, the peak-shaving value that solar provides in keeping a commercial building out of a higher-cost tier can be more valuable than the amount of baseline kWh produced by the array.

Energy prices also vary widely from one market to another. This affects the economic viability of solar projects. This in turn impacts the process for selecting one project in a given location over another project elsewhere. For example, across the United States electricity prices from state to state vary more than 500 percent from lowest to highest price. There is further variation from one utility territory to another within each state.

Not surprisingly, markets with high electricity prices tend to be large, dense coastal cities with high populations or geographic constraints. You can see intuitively that the price of electricity is an important driver in the feasibility of solar projects for commercial buildings. While incentives can compensate for lower energy prices, it certainly helps to have a solar project located where the energy it produces has a high value.

Quick tip

Grid electricity prices are typically highest during the summer, especially during the working week in the late afternoon, when energy demand spikes.

Technology

Solar equipment technology takes sunlight and converts it into usable electricity. The term "technology" is used broadly to refer to the solar module as well as the other electrical components that make up a solar array. The choice of technology is important

Figure 3.1 Solar modules being installed; a solar module being handled by two workers (Source: image courtesy of Greg Vojtko/US Navy, via Wikimedia Commons).

because it can affect the output of the array, its cost, and the impact it has on the underlying property.

Some solar module solutions may produce energy most efficiently in bright direct sunlight, but see performance drop off quickly in overcast conditions or if they are not tilted to directly face the sun. Highly efficient solar modules with an inefficient inverter and transformer will diminish energy output from the system. Other technologies may be highly efficient, but also expensive.

The way solar equipment is installed can affect system output. High-output modules installed flat on a roof may be no better than standard-performance modules installed at an optimum angle to capture the sun's energy. Many of these considerations will be reviewed in greater detail in subsequent chapters, but it is not difficult to see how technology plays an important role in the overall feasibility of a solar project.

Lessons from the field

Waiting for the next, best technology

I've spoken with a number of industry colleagues hesitant to move forward with a solar project because of a nagging worry that a few months after their array becomes operational, a new solar technology will come to market that makes their system obsolete. There are several reasons why this should not deter an otherwise attractive solar project.

Module efficiencies have increased by roughly 15 percentage points – but have taken the past 30 years to do so. That is an annual rate of 0.5–1 percent. New, "disruptive" solar technologies often grab headlines, but upon closer evaluation it turns out that the efficiency of these newer technologies lags traditional solar products. Advancements in the lab take years to reach the market, so you should not hold back for fear of a disruptive solar technology.

Furthermore, when solar projects are developed, the cost of the project along with the value of incentives and any financing are locked in based on the design and technology specified for the project. If the project makes economic sense, there is no need to wait. More importantly, while waiting for technology to advance each year, the incentives available to support your project may have declined. In the end, waiting for an improved technology could actually hurt your project's return if incentives shrink or disappear.

Finally, when new products come to market they usually command a premium price. So while the latest technology may allow your project to generate more electricity, you may pay disproportionately more for it. Installers may also face a learning curve when installing a new product if it is different to previous versions. This could drive installation costs higher for your project or even delay permitting if local code officials are not familiar with the new technology. These are the practical, real-world impacts that often outweigh the advantages that the latest technology may have on paper.

Each technology has its benefits and limitations, so there is no one single best technology solution for all projects. You may find, for instance, that a highly efficient solar module is the best fit for a space-constrained host property, while a less expensive and lower efficiency module is a more appropriate solution on a roof that has ample space for the array. The best solar projects are those that find the technology that balances cost, output, and building impacts, while taking into account insolation, electricity price, and incentives.

Incentives

Incentives are the most versatile ingredient in our recipe for creating economically viable solar projects. Because of their importance in supporting cost-effective financial solutions for solar, this chapter devotes a fair amount of space to describing their role and how they are commonly implemented in the marketplace. Incentives come in various shapes and sizes, and to maximize project value you may have to assemble multiple incentives from different sources.

In the United States, for example, project developers often pursue incentives from the federal government, the state government, the local utility, and even the local municipality if available. Outside the United States, incentives tend to be simpler and primarily based on a single predominant incentive provided directly or indirectly from national governments. As you will see throughout this book, capturing and optimizing the range of available incentives is a critical part of maximizing the economic value of solar projects.

Incentives are often thought of as cash subsidies, usually from the government or the regional utility. In reality, incentives can come in a variety of forms. Solar projects may be eligible for expedited permitting that allows you to reduce carrying costs. There could be a time-saving standardized process when applying to interconnect your project to the utility grid rather than the traditional time-consuming process used for other types of projects. Incentives for development projects may even come in the form of lower impact fees or even increases in allowable floor-to-area ratios in some markets.

There can be philosophical differences on the issue of incentives, and a range of opinions about what levels and types of incentives are appropriate. Fortunately, this handbook is not about policy, politics, or regulation. It accepts the world as it is today – good or bad – and explains how to make the most of what incentives are available to both you and your competitors. It operates under the simple assumption that if an economically beneficial incentive is available for your project, you will want to capture it.

For projects in most of the world, a FiT payment is the primary incentive solution available for commercial solar projects. This is a fixed unit price payment based on the quantity of energy delivered to the utility grid. FiT payments are typically expressed in terms such as "$0.20 per kilowatt-hour." The more energy produced by a solar array, the greater the FiT payment will be.

The incentive structure in the United States is different than in most other countries. Incentives are a combination of federal and state tax credits and accelerated depreciation. Added to that are utility incentives in the form of predetermined cash payments or production-based payments. These payments typically assume that the federal incentive can be captured and so they are set at a level that takes that into account. There can be other incentives from local municipalities, often in the form of expedited permitting

Quick tip

Solar incentives may include cash grants, tax benefits, and other forms of non-cash project assistance, such as streamlined permitting and expedited utility interconnection review.

or other non-financial project support. In today's world, your solar projects must be able to capture these incentives to maximize the value to the project.

In some cases, incentives can be more valuable than the amount of insolation a property receives. Germany, a country with an oversized share of overcast and cloudy weather, became the largest solar market in the world due in large part to its strong FiT program. Few other parts of the world, no matter how sunny they are, have yet to come close to catching up. Even in sun-drenched locations and with the continuing

Lessons from the field

Tax incentives: the US incentive model

The incentive landscape around the world is fairly consistent, consisting mainly of FiT programs. The most notable exception is the United States, where several layers of incentives must be captured for solar projects in order to maximize value. The incentive package for most solar projects begins with the federal tax credit and accelerated depreciation, referred to collectively in this handbook as *tax benefits*. The federal government provides a tax credit worth 30 percent of qualified solar project expenditures. The federal tax code also enables solar projects to qualify for Modified Accelerated Cost-Recovery System (MACRS; pronounced "makers") depreciation, to recover investments in solar property more quickly. Solar projects are eligible for 50 percent bonus depreciation. Together, the tax credit and accelerated depreciation comprise the single largest incentive for projects in the United States. These incentives are currently set to expire at the end of the year 2016.

Many commercial property owners considering solar face the challenge of having have a limited tax appetite due to the structure of their real estate investments. This can create complexity in the solar project because a third party investor becomes necessary to monetize the solar project's tax benefits. In these cases the tax credit and accelerated depreciation are bundled and transferred to the third party investor in exchange for an up-front cash payment. This unlocks the value of the tax benefits for the project sponsor. Many solar companies offer tax equity financing solutions to capture the tax benefits of a project regardless of the property owner's tax situation. Regardless of whether you can use the tax benefits yourself, they are a significant portion of solar project value. Summaries of federal renewable energy tax incentives can be found at: www.dsireusa.org.

improvements in technology performance and cost, incentives still have a role to play in maximizing the financial performance of solar projects on commercial buildings. A key takeaway is that the opportunity for financially viable solar projects is not limited to the sunniest markets or those with the highest electricity prices.

Utility incentive programs

Solar incentive programs are typically mandated by policies created by national or regional governments, but the programs themselves are often implemented through local or regional electric utilities. The utilities are instructed to administer a solar incentive program, including the payments that support solar projects. Utility incentive programs can be a primary source of incentives in some markets, while in others they may have a small value compared to tax benefits or other incentives awarded outside of the utility. Some utilities may offer lump sum cash grants to qualifying solar projects, while others may fund project incentives based on the electrical output of the system over several years. Others have set up trading markets that place a value on the clean energy attributes of solar electricity to enable project owners to sell these attributes in order to offset project costs. These attributes are known as renewable energy certificates (RECs) and are bought and sold in an open marketplace.

Having more than one utility in a real estate market can add a wrinkle for property owners interested in solar. Utility markets may have one dominant utility, but small municipal electric utilities may provide electric service in certain areas within the primary

---

Lessons from the field

Why utility territory matters

California illustrates how utility territory can affect whether or not a project qualifies for incentives. California has a handful of very large public utilities that supply power to the majority of the state's commercial properties. There are, however, dozens of other small municipal electric utilities operating in the state.

Take, for example, the extreme case of the City of Vernon (population 113; 5.2 square miles). Vernon, a small city with few residents, where the real estate stock is composed primarily of warehousing and manufacturing uses, is surrounded by the City of Los Angeles (population 3.8 million; 503 square miles) and adjacent satellite cities. Immediately to Vernon's west and north solar incentive programs are administered through the Los Angeles Department of Water and Power. Properties to the south and east of Vernon lie within the Southern California Edison utility and solar program territory. At the time of writing this book, Vernon did not have a solar incentive program, so properties receiving power from the City of Vernon Light and Power do not have access to solar incentives that commercial property for miles in every direction can tap into. While the sun shines on Vernon as it does nearby, the lack of incentives makes solar a hard sell in this small industrial city.

utility's territory. In many cases they have different regulatory requirements when it comes to solar incentive programs. They may not even be required to have solar incentives at all if they fall outside the regulated group of utilities that the regional solar policy applies to. This is a particularly common challenge in the United States, where there are thousands of these small municipal and private utilities in both small and large cities. Because of this patchwork of utility territories, it is possible that buildings near one another in the same city could have drastically different solar incentive programs.

Where there is more than one utility solar incentive program in a property market, you should recognize that the programs might not support projects of the same size and scope. Incentive programs are often capped at a certain size, either by limiting the dollar amounts available for incentives, or by establishing a first-come, first-served queue of projects up to a certain pre-defined cap. Many utilities have programs that target small commercial installations, while a smaller subset also offer programs that support large projects. There is no universal rule for what qualifies as a small, medium, or large size project. In general, for commercial buildings this handbook defines project scale as:

*   small: less than 50 kW;
*   medium: 50–250 kW;
*   large: greater than 250 kW.

Utility programs come in a wide range of shapes and sizes. Some programs may specify a total dollar amount that they have available to support projects that they disburse on a first-come, first-served basis. Other programs may have a program size target in megawatts, and will fund projects until they reach their cap. Within these programs, there will often be limits on qualifying projects based on size, cost, and other factors. These restrictions can either be a benefit or a frustration.

For example, if you want to offset a large portion of your electricity use with solar, you may find that a program that restricts projects to less than 50 kW impairs your ability to achieve meaningful progress toward your goal. In this type of market, solar is likely to create a small amount of savings in relation to your building's operating budget. Conversely, if you want to support the efforts of a small tenant that wants to "go green" and offset a portion of their energy use, a program of this size may be a good fit.

At the other end of the spectrum are programs targeting large-scale projects. For these sizable systems, unless there is significant energy demand on-site or a mechanism to sell power to utilities, many commercial properties will find themselves unable to justify such a large project.

Utility programs may also restrict funding to certain locations or property types by earmarking incentive funds to target areas such as economic improvement zones, brownfield sites, or affordable housing projects. This can be advantageous for qualifying properties, but it excludes other property owners entirely. The state of New Jersey has deployed solar incentives in this manner; for example by earmarking funds for projects in designated urban enterprise zones. These specialized program blocks tend to be relatively small – maybe only a few million dollars and a few megawatts in total – and are easily filled by market demand.

Programs that target a small subset of the commercial market can be beneficial for the utility, capping their financial obligations and allowing them to quickly fill program

Lessons from the field

Capacity vs. production-based incentives

There are two common types of incentive payment categories. Utility or government incentive programs typically offer payments based on one of the two following measures:

1 installed capacity;
2 electrical output.

These are referred to, respectively, as capacity-based and production-based incentives. Capacity-based incentive payments are determined by the nameplate rating of the solar array, measured in kilowatts. Capacity-based incentives tend to be more common for smaller solar facilities, such as those on residential and small commercial properties where ongoing monitoring of output would be disproportionately costly or time-consuming compared to the size of the array. Capacity-based incentives do not take into account the quantity of electricity that is produced.

In a production-based incentive, payments are tied to the energy output from the array. Incentives are then paid periodically based on the total kilowatt-hours produced. Production-based incentives are more common for commercial-scale projects than for small-scale projects, in part because they represent a larger investment in renewable energy and also because it is more cost-effective to monitor these larger solar facilities. FiTs are a form of production-based incentive.

Under a capacity-based grant program a 100 kW-size solar facility would receive a greater cash grant than a 90 kW solar facility, even if the 90 kW system produced more electricity each year. The opposite would be true under a production-based incentive program.

quotas. However, it does little to support consistent and widespread solar project planning and adoption. It also promotes a "gold rush" mentality within the solar market where project developers flood each program with applications, hoping to be one of the few selected. This is not only inefficient; it limits the ability of the solar industry to develop in a rational manner over the long term.

Just as importantly, this behavior largely prevents commercial property owners from pursuing solar projects unless they are already queued up to jump on fast-moving opportunities. Short, aggressive, quota-filling program timetables are often at odds with institutional ownership of commercial properties. These properties generally require investor and lender approval, or at a minimum a courtesy notification, for an undertaking such as solar. The amount of time this takes may cause the project to lose out on the opportunity to participate in small-quota programs. There are solutions to overcome this incompatibility, but these incentive structures tend to discourage property owners from participating in short-lived solar programs.

Quick tip

Look for incentive programs that support long-term stability and scalable growth where the solar market can expand over time. Be wary of small-quota programs that are likely to reach capacity very quickly.

As noted previously, regulators have developed incentive programs that are designed to monetize the environmental attributes of solar electricity. This is accomplished by creating a market where projects can sell RECs generated by solar projects. Buyers in this market are often utility companies that have a regulatory mandate to purchase renewable energy certificates equivalent to a certain percentage of their energy needs. By adjusting the utility company's demand for RECs, regulators can influence the market price, and thereby the incentive available to support solar projects. Because the utility companies are obligated to buy RECs in this marketplace, they provide the demand for the development of solar projects.

The value of RECs can vary widely depending on the market where your property is located, and in markets that lack a mechanism to trade RECs, their value is virtually zero. For example, the REC trading market in New Jersey has functioned as the primary source of project funding for renewable energy projects in the state. The sale of RECs in this market has in the past translated into an incentive worth as much or more than even the federal tax credit. More recently, the value of RECs declined because of reduced buyer demand.

Regional and local incentives

The most valuable incentive programs are typically available from the federal level or from utilities administering renewable energy programs, but there can also be incremental incentives that are made available at the regional and local level. Check the availability of incentives offered by the county or city where your project is located. These incentives tend to be less valuable than primary federal incentives but they may nonetheless be beneficial enough to pursue. Common types of state incentives include:

- property tax exemptions for solar property;
- sales tax exemptions on solar materials;
- accelerated depreciation for solar equipment;
- streamlined permitting and zoning exemptions;
- cash grants for projects that use the local labor force.

These incentives can vary widely, and may be entirely absent in some markets. Some will be worth pursuing, and others may have little value for commercial projects. Some may not be worth the hassle of pursuing due to paperwork compliance requirements or restrictive rules that could actually increase project costs, such as a local hiring rule or a buy-local solar equipment purchase requirement. Because these incentives are typically less valuable to the project, failure to capture them is unlikely to derail your solar project. They can nonetheless add value and should be considered.

Municipalities tend to be less likely to provide cash grants or tax benefits, but they are more likely to provide incentives in the form of development support. A municipality may offer expedited permitting for solar projects to speed up the design review process. They may relax certain zoning rules, such as reducing setback requirements for solar equipment. In other cases, historic district rules may be amended to allow solar projects on buildings where they would have otherwise been prohibited.

Where cash incentives are offered, they tend to be limited in value, and may come with strings attached. Check what the requirements are to comply with the grant's eligibility rules and compare them to the value of your project. It is not uncommon for grants to come with additional compliance requirements such as:

- paperwork, including engineer-stamped and utility pre-approved application;
- pre- and post-construction inspections;
- access to solar array performance data;
- local hiring rules;
- consent to audit project;
- union-scale wage rate labor requirements.

These additional costs should be considered alongside the potential value of the incentive in order to determine if applying for it would really bring value to the project. As an

---

Lessons from the field

Net metering

Net metering is an essential energy policy that enables solar projects to be connected to the electric utility distribution network, stipulating that utility companies must allow any excess energy that customers generate to flow back to the grid. This is what is referred to colloquially as "spinning the meter backwards." Net metering policy opens the door to enable solar projects to be installed on commercial buildings.

While utilities subject to net metering are required to allow excess power to be returned to the grid, they may not be obligated to pay you for it. In the United States many utilities offer a credit against future usage in lieu of cash payments. If they do compensate the project owner, it is often at a low wholesale cost of power. This may be a few cents per kWh, a fraction of the retail rate paid by customers. Contrast this with markets that have FiT programs, where the payments from utilities often equal or exceed the price of electricity.

Fortunately, net metering programs exist in many countries. Within the United States, 40 states and the District of Columbia have net metering rules. As a result, the likelihood of your project being located in a market with net metering is high. If you've seen solar on any rooftops at all in your area, you are probably in a market that has net metering. Solar companies or utility representatives can quickly tell you whether a particular market has the necessary net metering policies in place to support solar projects.

example, in California the city of San Francisco has provided a cash grant program that offers up to $10,000 per project. This grant includes many of the requirements listed above. For a project that already has some of these requirements, this may not be a big deal, and pursuing this incentive may be justified. For a project that would otherwise not be subject to these requirements, the value of the cash rebate may be largely negated by the cost of compliance. For a particularly large project, an incentive of this size may not have much of an impact on the project's bottom line, particularly once the compliance requirements are included.

Researching incentives

When you want to keep up with incentive programs in more than one market, there are several resources you can look to. Solar companies that have executed projects in the markets where you are focused can help sort through the incentive programs. They can quickly identify where in your portfolio the most viable solar projects exist – they do not want to waste their time in poor solar markets any more than you want to waste yours. These initial screening exercises cost you nothing – they are as much for the solar company to understand what opportunity you may present as they educate you and try to win your business.

Real estate service providers such as CB Richard Ellis, Jones Lang LaSalle, and others like them have solar advisory practices that can support your solar efforts. Engineering firms, solar consultants, solar developers, and even some solar product manufacturers can also help you review solar incentives in a given market. For purposes of the handbook, we refer to the real estate service providers, consultants and advisors collectively as "solar service providers." This is a broad category for the companies that can provide solar industry knowledge and project assistance. We will discuss solar service providers in greater detail in Chapter 5.

As of the writing of this handbook, most Western and Central European countries offered solar incentive programs. Portions of Canada and almost half of the states in the United States offer commercial solar project incentives. Japan re-launched an incentive program in 2012 and China has begun pursuing solar, although it is not actively targeting projects on commercial buildings yet. The countries and regions offering incentives can and do change frequently.

Within the United States the most active states for commercial building projects have been California, New Jersey, Colorado, Massachusetts, and New York. More recently states such as Pennsylvania, Arizona, and Texas have introduced solar programs that support solar on commercial properties. This is not a definitive list, however, and other markets could become active at any time. Electric utilities publish information about their solar programs on their websites, listed under headings for renewable energy, environ-ment, or under business services. Be careful to note whether solar programs are targeting commercial buildings, homeowners, or other types of projects.

Final thoughts

This chapter explained the four elements of economically viable solar projects. Insolation is a function of location that affects the amount of energy your solar array is capable of producing. Electricity prices affect the value of the kWh produced by your solar array.

This is directly applicable if you are offsetting the energy consumed by a building, and indirectly where it is a factor in determining the rates paid by the local utility for excess electricity. The technology that goes into the solar array affects the cost and energy production of a project as well as its feasibility to be installed on buildings. Incentives of all types – whether federal, utility, or local – can have a significant impact on project economics.

The diverse range of incentive programs can be daunting for many property owners to sort out. Fortunately you do not need to become a solar expert to find out what incentives are available at your property's location. Completing the portfolio assessment that you find in Chapter 4 will provide you with the information to begin discussions with solar companies, consultants, and real estate service providers that can stay abreast of incentive programs on your behalf. Using these companies as a resource will help you find out what options are available for your property without having to spend the time researching programs yourself.

Each of the four ingredients of a successful solar project can be used in different quantities, depending on where your project is located and how it is financed. It is important to find the best opportunities in your portfolio that allow you to create a project that optimizes each of them. This will result in the most economically advantageous project for your needs. The following chapter explores how you can quickly and reliably find opportunities within your property portfolio that allow you to make the most of the combination of ingredients that are available in your target markets.

Chapter 4

Identifying opportunities in your portfolio

Chapter summary

- A portfolio-wide property review allows you to rank properties according to their likelihood of supporting successful solar projects.
- Gathering portfolio property data allows you to quickly re-assess the feasibility of properties as the solar market changes over time.
- Data gathering can be divided into five categories: location, asset management, building, tenants, and energy.
- Target properties in locations that have strong incentive programs and insolation levels.
- Review asset management plans to identify properties that may be poor candidates due to planned sale, redevelopment, and other asset-related considerations.
- Target properties that have physical building characteristics that support solar.

This chapter describes a process to review the suitability of pursuing solar projects at the buildings in your commercial property portfolio. The process takes into account a range of physical and operational factors that impact solar project feasibility. It also identifies how to find out if the essential building blocks from Chapter 3 are available in a given location. This will allow you to quickly prioritize the properties where your efforts will be most likely to succeed, while avoiding wasted time on properties that have little chance of working out. The portfolio review process identifies "must have" features that are essential to even consider pursuing a solar project.

The remaining characteristics of your properties will be used to allow you to screen the portfolio for solar opportunities. For example, you may have a subset of properties that are good candidates for one solar incentive program, and other properties that are better suited to another due to location or roof area. There are numerous factors to consider, and one of the most time-efficient ways to determine feasibility is to share your completed portfolio review with a few service provider companies in the solar industry, such as solar developers. Ask them to rank the properties and tell you why certain factors mattered and why others did not for a given market or property.

Between compiling the review list and getting feedback from solar companies you will be able to develop a sense of where the strongest opportunities lie. As the solar landscape evolves you will also be in a good position to reassess properties and quickly determine

Quick tip

Net metering is a type of legislation that enables solar electricity to be returned to the electrical grid. Net metering is a necessary feature of a viable commercial solar market.

if solar market changes turn weak sites into strong ones. In the end, you will be able to determine where in your portfolio it makes the most sense to pursue solar projects.

Step 1: gather property information

The first step in identifying the most promising solar candidate properties is to gather a range of information about your properties. This process identifies key operational and physical factors that you will need to understand before proceeding to more time-intensive levels of due diligence. It will quickly tell you where to devote more time – and where to avoid wasting time – so you can identify properties that have the highest likelihood of success.

Collecting this information also enables you to talk to service providers and others in the solar value chain about opportunities that may exist in your portfolio. This information will also come in handy when it is time to design a solar array, apply for incentives, and seek project financing. The review can be performed at one of three levels: (1) for your entire property portfolio; (2) for a subset of related properties; or (3) for a single asset. The advantages and disadvantages of each are described in Table 4.1.

Conducting due diligence at the portfolio-wide level is generally more time-efficient over the long term compared with doing one or a few assets and then expanding incrementally to include more. Reviewing your entire portfolio allows you to identify

Table 4.1 Levels of portfolio review

Assessment level	Advantages	Disadvantages
Entire property portfolio	• Identifies scope and scale of solar opportunity • Saves time vs. ad hoc assessments • Simplifies sharing information with solar companies	• More time-consuming initially than smaller-scale assessments
Subset of related properties	• Targets specific asset types or specific markets • Selects for similar features desirable for solar • Saves time targeting properties in known solar markets	• Shared characteristics may limit solar opportunity for entire group • Opportunities outside of targeted properties may be overlooked
Single asset	• Allows highly detailed assessment for a specific opportunity • Targets property with high value for solar	• Adding other properties requires due diligence to be repeated • Failure of chosen property requires re-start of due diligence

the overall scope and scale of your solar opportunity. By doing this you will have a sense of whether solar is a small-value opportunity available to only a few properties, or if it is a multi-million-dollar opportunity that may warrant a much more systematic approach to developing the business case for a focused solar investment program. In the same way, this high-level review will help you decide if it is worthwhile allocating staff time to develop permanent solar expertise for a portfolio of projects, or if it is better to leave the few opportunities in the hands of third-party solar service providers.

A portfolio-wide review saves you time over the long run, compared to ad hoc assessments on individual sites. Instead of guessing which property might support solar only to realize that the opportunity lies elsewhere, you can identify the best market and tackle your properties in that market in a systematic way. You will also be able to eliminate a significant amount of redundancy when talking to lenders, joint venture partners, utilities, and other stakeholders to gather information. For example, it is much faster to provide a single complete list of properties to a lender or utility at one time than it is to ask them individually for the same information for each property.

Table 4.2 illustrates an example portfolio along with a basic comparison of the building blocks of solar projects to demonstrate the value in performing a portfolio-wide assessment. You can see that Building A is in a location where the basic ingredients of solar projects are available – insolation, high electricity prices, and valuable incentives. You can also see that Building B may be a good candidate, except that the available area for solar is quite a bit smaller than Building A. Building C might appear to be an even better candidate than A or B, given the large roof area, except that it is located where energy prices are low and the solar incentive program has little value. While Building D has low insolation and moderate electricity prices, it is likely to turn out to be a better candidate than C due to the existence of a strong incentive program. And while Building E has a large available area, high insolation, and moderate energy prices, the lack of an incentive program prevents it from being considered in today's solar market.

Having information about your property portfolio in one place prepares you for conversations with solar companies and service providers. In order for them to provide input that helps you identify the best solar sites, they will request some or all of the information that you gathered in your due diligence process. You can advance the conversation with these companies more quickly and collaboratively by providing the information upfront which they will ask you for anyway. Making the portfolio-wide opportunity known to these companies also provides greater incentive for them to want

Table 4.2 An example portfolio

Property	Available area (square feet)	Location	Insolation	Electricity price	Incentive value per kW
Building A	50,000	San Diego, CA, USA	High	High	High
Building B	15,000	San Diego, CA, USA	High	High	High
Building C	90,000	Dallas, Texas, USA	High	Low	Low
Building D	55,000	Toronto, Ontario, Canada	Low	Medium	High
Building E	70,000	Monterey, Mexico	High	Medium	None

to work with you because they can see the value of building a long-term relationship rather than just focusing on maximizing their profit on a single project.

Portfolio-wide due diligence is a way to manage time and risk. If one property is determined to be unsuitable for solar, you will lose little time transitioning to another property that has already been pre-screened. In the fast-moving solar industry, this helps avoid missing opportunities while you hunt for new host properties. Looking at your portfolio as a whole also provides insight into the characteristics of your properties that will become useful for future opportunities as solar incentives and technologies evolve. Just as importantly, you will be able to see the economics of solar at scale rather than as a series of one-off installations.

As described above, there are compelling reasons to complete a portfolio-wide assessment, but there are also times when analyzing a limited subset of your portfolio may make sense. This could occur when sites are co-located in a known active solar market, such as a utility territory or even a country that has solar incentives. They may be assets of a single property type that tend to be good solar candidates, such as distribution centers and warehouses that have large rooftops. There could also be common ownership structures such as a business park or a campus where solar can be deployed across adjacent properties. While these considerations can also be picked up in the portfolio-wide assessment, time constraints or other resource limitations may require a more targeted approach that focuses on a smaller portion of the portfolio.

Table 4.3 illustrates several ways a limited portfolio due diligence process can be valuable. If you have a geographically concentrated portfolio, this is the scenario you will face. Alternatively, you may want to focus on a single market because you have already determined that the market has high levels of insolation and high electricity prices. Starting with Building 1, it is apparent that while it is in the same geographical market, the lack of an incentive program makes this property unattractive for solar. Building 2 falls within the incentive program size range. The roof of the building is old, but assuming the solar array can be installed at the time the roof is replaced makes this a promising candidate. Building 3 already has a new roof, but it is too small to qualify for the incentive program. Building 4 appears to be a good candidate except that the roof is mid-life. This means that the roof will need to be replaced sooner than planned, or after the solar array is installed. In either case there will be additional costs to the project to address the roof. Building 5 appears to be an excellent candidate for solar. The only caveat is that the building can support a solar array many times larger

Table 4.3 Limited portfolio due diligence

Property	Available area (square feet)	Location	Incentive program value per kW	Incentive program size range (equivalent square feet)	Roof condition
Building 1	90,000	City of Vernon CA, USA	None	None	New
Building 2	75,000	San Diego, CA, USA	High	50,000 to 100,000	Old
Building 3	15,000	Los Angeles, CA, USA	High	50,000 to 100,000	New
Building 4	60,000	Rialto, CA, USA	High	50,000 to 100,000	Mid-life
Building 5	275,000	Riverside, CA, USA	High	10,000 to 50,000	New

than the solar incentive program will fund. This will leave a large portion of the roof under-utilized. While this may not be a problem per se, it does limit the value that can be obtained from installing solar on such a large building.

It may also be appropriate to consider only a single property if you already have knowledge of the solar market, or if you already have a good sense of which properties will be more viable. Looking at single properties is more likely to make sense if you are looking only for a pilot project or if you have solar in mind to achieve a specific purpose. Reviewing a single asset could also make sense if you are jointly exploring solar at the request of a building tenant. Time constraints could also lead you to choose a single project, although by the time you identify a building that fully supports the solar opportunity, you may have ended up repeating the data collection effort several times for several buildings.

You may also consider solar for a single property if the intended solar design is highly customized for a particular building and therefore unlikely to be replicable at another site. This could be a project that you are planning to redevelop where adding solar is part of the project's goals to earn a sustainable building certification or if solar helps you successfully complete city and community reviews.

Quick tip

A due diligence spreadsheet allows you to quickly screen your portfolio as new solar markets emerge or as incentive programs change over time. Once the spreadsheet has been created, it is easy to keep it updated as your portfolio changes over time.

Due diligence categories

To begin the property due diligence review, create a *solar due diligence* spreadsheet that contains a list of your properties, organized by market. Group the data in the spreadsheet into five general headings:

*   Location.
*   Asset management.
*   Building.
*   Tenants.
*   Energy.

The first two categories are populated primarily with the critical factors that drive the viability of your solar projects. The remaining categories are factors that generally influence the cost and design of a project that is otherwise viable. In order to populate the spreadsheet with data for your properties you may need information from several sources such as a property manager, asset manager, and tenant. Create columns for each sub-heading listed below, and put each property in a row. Begin filling in the cells for each subheading that corresponds to each property. Under the *location* heading, as illustrated in Table 4.4, include columns for the following property location characteristics:

- Property name.
- Street address.
- City and state.

For the *asset management* heading, add subheadings named according to the following inputs:

- Property ownership entity.
- Debt holder, if any.
- Candidate for redevelopment or sale (yes/no), and when.

At this point you should have a spreadsheet that looks like Table 4.5.

Continue adding columns to the spreadsheet to further expand on Table 4.5. For the *building* category, add subheadings named according to the following inputs:

- gross roof area;
- year constructed;
- roof age;
- remaining expected roof life;
- type of roof (e.g., "EPDM," "TPO");
- roof warranty: original term and years remaining (e.g., "20 years / 7 years");
- height of roof (approximate height above ground level);
- amount of rooftop equipment, including skylights (e.g., "minimal," "moderate," "extensive").

Table 4.4 Location due diligence categories

Location			
Property name	Street address	City	State
Building 1	123 Main St.	Los Angeles	CA
Building 2	456 Main St.	Boston	MA

Table 4.5 Asset management due diligence categories

Location				Asset Management			
Property name	Street address	City	State	Ownership entity	Debt holder	Redevelop-ment or sale	When?
Building 1	123 Main St.	Los Angeles	CA	SoCal Properties, LLC	Bank A	No	n/a
Building 2	456 Main St.	Boston	MA	Beantown Partners, L.P.	n/a	Yes	January, 2014

In the same manner, continue to add columns for *tenant* characteristics:

- Tenant name.
- Contact information.
- Lease expiration (e.g., "9/6/2015").
- Lease type (e.g., "triple-net," "full service gross").
- Party responsible for roof maintenance (e.g., "owner," "tenant").

Add columns for *energy* use:

- Name of electric utility provider.
- Annual total energy use for landlord, in kWh.
- Annual total energy use for each tenant, in kWh.
- Retail price of electricity in \$/kWh.
- Utility rate schedule, if known.

This may seem like a lot of data to gather, but it will be needed in order to determine which properties are the best solar candidates. It is okay if you do not have data for each and every category in the portfolio review spreadsheet. Gathering as much of this information as possible at the outset will avoid delaying subsequent due diligence steps. Once you have completed the review, this data will provide a sufficiently complete picture of your properties to move to the next step and begin prioritizing them for solar through an initial screening process.

Lessons from the field

Due diligence shortcuts

Some solar companies determine a property's clear roof area with an online aerial satellite mapping program such as Google Earth, then ask how much of the roof warranty remains, and call this initial due diligence. This approach may be adequate to get you to an indicative proposal quickly, but it lacks the detail necessary to ensure a successful project. It ignores important due diligence considerations that will ultimately be critical to the project and that can result in a great deal of wasted effort. To protect your properties and avoid wasting time, look for solar companies that are upfront about the type of due diligence that will be needed.

While they may provide an initial indicative proposal, they should ask about the screening factors covered in this handbook as part of a complete and thorough due diligence process. Addressing these necessary due diligence considerations will allow you to identify the best sites for solar projects more quickly and reliably.

Step 2: initial screening

Now that basic property data is collected in one spreadsheet, you can begin analyzing your portfolio. Conducting an initial screening of your portfolio has one major goal: to allow you to identify any properties that have flaws that prevent solar from being feasible. Identifying these properties quickly at this stage saves you from unnecessary time-consuming due diligence just to learn that the asset in question is not suitable. This could be perhaps because the property is located in an area that lacks solar incentives, or because the property will be sold in the foreseeable future before a solar project could be completed. The initial screening of your portfolio assessment looks primarily at the first two categories from the due diligence spreadsheet: *location* and *asset management*.

Location

It is helpful to begin with location because this determines both insolation and incentive eligibility – two of the four ingredients necessary for a successful solar project. At this stage, if you know solar companies or other solar service providers you can share your spreadsheet with them. They can help identify incentives where the properties are located, along with relative levels of insolation. Solar companies can tell you what combined effect these have on project feasibility. If a solar company is not available, do not worry. Review the asset management sections and eliminate any poor candidates due to those factors. You may find it easiest to send a solar company your due diligence spreadsheet and ask them to add columns and fill in specific solar market data.

Asset management

In addition to location, consider your asset management objectives and decide if solar fits in with them. Talk to the people in your organization that have a say in the matter – executives, fund managers, development managers, joint venture partners, and the like. Be on the lookout for feedback that either explicitly or implicitly suggests that making a long-term investment in solar could lead to conflicts with their asset management plans. Examples of asset management considerations include:

- plans to sell the asset before the solar project would be completed;
- adding debt or refinancing debt that would subject the solar project to lender consent;
- redevelopment plans that would impact the area where solar is planned;
- capital improvement projects that would interfere with solar project construction.

Note this feedback in the due diligence spreadsheet. Include any time-sensitive considerations if the situation is expected to change in the future. If there are asset management plans that preclude solar that are unlikely to change, remove that property from the due diligence list. For example, you will save time and aggravation by identifying and eliminating properties slated for redevelopment a few years after your solar project would have been completed. Conversely, a near-term roof replacement may provide a perfect time to incorporate solar. Identifying these factors will support more productive conversations with solar companies, service providers, and consultants in subsequent phases of due diligence.

Ownership entity

The ownership structure of a property may impact its feasibility to pursue solar. Decision-making for a wholly owned asset will be simpler than for a property where you also need approval from one or more partners. For a property held in a joint venture (JV), obtaining partner approval could add complexity to a solar project. The JV needs to be able to address the solar deal effectively in the event that there are costs associated with the solar project, or if the joint venture is unwound in the future. Properties held in co-investment funds may also require approval from lead investors prior to implementing solar projects.

The legal structure of project ownership is often overlooked by solar companies in the initial stages of due diligence, but it can often be a critical factor in determining what options the property owner has for pursuing solar projects. For example, many solar companies initially underwrite projects assuming the property owner can obtain the full value of all available tax credits. However, many commercial properties are owned by entities that have little or no tax liability, so this is not feasible. Other solar companies may be surprised to learn that a proposed solar project requires approval from fund investors or a JV partner.

As noted, the type of ownership structure that controls the property can also have an effect on eligibility for certain incentives such as the federal tax credits. For example, in the United States real estate investment trusts do not pay corporate tax and therefore cannot readily claim the federal investment tax credit. This may also be the case for

---

Lessons from the field

Other perspectives on solar

I have found that tenants, lenders, and potential acquirers often take a pre-existing solar array on a building in stride. More often than not, tenants appreciate the efforts you have made to operate responsibly, whether or not there is a direct benefit to the tenant. It is still prudent to consider the types of tenants that currently, or in the future could, lease space in the building. Will they be attracted to a building that has solar? Or will they be concerned about the solar array increasing maintenance costs or restricting their use of the building? For the landlord, the solar array can serve as a differentiator in a competitive marketplace.

Several lenders that I have dealt with view solar positively because they can point to it as a way in which they are supporting renewable energy. This is not a universal view, but it does show that lenders are considering it. Others in the real estate industry, particularly institutional investors looking to acquire properties, are likely to point on one hand to the solar array as a source of higher maintenance costs, while on the other hand they are sure to be adding the solar revenue to their financial analysis. Not surprisingly, the sentiment on solar can be mixed. This makes it even more important to be able to communicate clearly and deliver a project that creates economic value while minimizing risks to all parties involved in the project.

properties owned through a partnership or a limited liability corporation. These considerations may not prevent you from pursuing solar projects, but they are likely to affect the type of solar project financing models that will be cost effective. This will be discussed in greater detail in Part III of the handbook.

Debt

Having debt on a property is usually not a problem for deploying solar projects, but it is a consideration that should be addressed to avoid surprises that could slow down the project or add an unexpected layer of sign-offs. Review loan documents and talk to the lenders to verify what input into the solar approval process they need to have. It may make little sense, but you may find that a lender hesitates when asked to approve adding a solar array to a property where they already hold debt, but they do not have a problem providing a loan to a property that already has solar on it.

Step 3: additional screening

As you complete the initial screening you will begin to have clarity on which assets are worth pursuing further when it comes to asset management considerations, and perhaps location if you have spoken to a few solar companies. For the additional screening, you need to look at other factors that come into play when deciding whether a particular property is a good candidate for solar. This additional screening of your property assessment list looks at the following characteristics:

- buildings;
- tenants;
- energy.

This additional screening is valuable to complete, whether you have shared your spreadsheet with a solar company or not. If you have, it is quite likely that they may have already come back to you requesting some or all of this additional detail. In terms of importance, these factors range from "extremely important" to "nice to have," but they are much less likely to derail a solar project completely than asset management or location factors. They are important to identify though, because they can have an impact on the design, costs, and administrative requirements of the project. Taken as a whole, these are factors that will enable you to rank your properties from the most to the least attractive in terms of where to deploy solar.

Buildings

Recall from earlier in the chapter the following list of building characteristics, listed below, that you have already added to your due diligence spreadsheet. These factors provide an indication of the general suitability of the building itself as a host for solar. The paragraphs that follow explain why this type of data is needed. Similar considerations will also apply whether solar is planned for areas of the building such as the roof, façade, or an attached canopy. Because rooftops are often the preferred location for solar arrays, the list includes several questions focused on roofs. Feel free to modify or skip this section of the due diligence spreadsheet if you are certain you are only looking at non-roof solar arrays, such as a freestanding parking lot canopy or a façade project:

* gross roof area (or parking lot, or façade area);
* building age;
* roof age and remaining expected roof life;
* roof type;
* roof warranty: original term and years remaining;
* roof height;
* rooftop equipment, including skylights.

Gross roof area

For many commercial buildings the roof is the preferred space for hosting a solar array. The area available for the solar array must be considered from the outset because it impacts how much electricity can be generated by a solar array. Incentive programs also often explicitly or implicitly target projects of a certain size. This translates directly to the area needed to host the solar array. If your property does not have the space needed for a project of that capacity, you will not be able to build a project that qualifies for incentives. This can affect both large and small projects. At one end of the spectrum, a project may not meet the minimum size threshold required to be eligible for the solar incentive program. At the other end, buildings suitable for large solar arrays may find that incentives are intended for small projects and have little relative value for large and costly solar installations. The same thinking applies to projects that are to be located on the ground as a parking canopy or on the façade of the building. Knowing the available area helps you determine how feasible the site is for solar.

For example, in order to sell electricity back to the utility provider under an independent power producer program in southern California you would need a roof area of approximately 100,000 square feet or more in order to meet the minimum size requirements for this program. Compare that to a typical office building, which has a floor plate of 60,000 square feet – a building with a roof of this size would be too small to participate in this program. At the other end of the spectrum, the city of San Francisco's solar incentive for commercial properties has been capped at $10,000. For a solar array covering 100,000 square feet this incentive would be worth less than 1 percent of the project's total cost.

Building age

Building age comes into play in two ways when assessing properties for solar. First, it may be an indicator suggesting that an asset is due for redevelopment in the foreseeable future. This possibility should be captured in the asset management review section of the due diligence spreadsheet. Second, and more importantly, building age can be an indicator of the structural capability of the building to host solar. The ability of this existing structural system to support what can be a significant additional weight from a solar array can impose restrictions on the weight and therefore the size of the project.

It is important to recognize the conditions where this is more likely to occur so it can be investigated properly during the due diligence process.

A solar array adds weight to the building, which it must be able to safely carry; this often means complying with today's structural and seismic codes. Because these codes tend to become more stringent over time, the added weight of solar equipment may push older buildings beyond their safe code-prescribed limits. Installing solar on outdated buildings may create a safety risk that can only be remedied by retrofitting and enhancing the building structure. This adds a financial burden to the project and potentially disrupts tenants in the building. More stringent codes are particularly likely in earthquake-prone regions where there is also widespread solar activity, such as Japan and the western coast of the United States. Structural codes have also evolved significantly in storm-prone coastal regions where hurricane and strong wind forces have been shown to affect building integrity.

While older buildings are generally assumed to be less suitable for solar because of code changes over time, there can be exceptions. Much older buildings may have been more robustly built due to construction practices in place at the time, or to accommodate building systems in use at the time that were larger and heavier than those used today. This can be seen with a roof structure that was originally designed to support a heavy gravel ballasted roof system. Removing a roof system such as this that can weigh as much as six pounds per square foot and replacing it with a modern membrane roof weighing two pounds per square foot would make available significant structural capacity. The additional capacity could then be used to support the weight of a solar array.

In extreme cases structural limitations can make a building unsuitable for solar. More likely, however, is that it will affect the layout, design, and choice of technology for the array. This could result in a project significantly smaller than originally planned, or increase costs beyond what the project can bear. It is essential to work with a structural engineer to evaluate the structure of the building because its age can – for better and for worse – affect the ability to support the added weight of a solar project.

Lessons from the field

Solar incentive resources

You may be eager to develop a basic knowledge of solar incentive programs to help with due diligence. A well-known resource for identifying solar incentives based on location in the United States is the Database of State Incentives for Renewable Energy (DSIRE). This site hosts a comprehensive set of information on state, local, utility, and selected federal renewable energy programs. You will likely find an overwhelming quantity of information at the site. Fortunately, it is not necessary to spend a great deal of time sorting through it all. You can use the knowledge of solar service providers to complete this screening process for you when the time comes. The DSIRE site can be found at: www.dsireusa.org

Roof age and expected roof life

The age and quality of the building's roof is a factor that has a direct impact on how suitable a building will be for solar. A roof that needs to be replaced in the middle of a solar array's lifetime is a costly and disruptive event – one that is best avoided. Solar projects are designed and underwritten with an expected useful life of 20–25 years, although they can operate for longer when well maintained. In comparison, commercial buildings typically have roofs expected to last anywhere from 15 to 25 years, but few were planned with solar in mind. It is important to understand where your rooftops are in their lifecycle so you can decide how to address their condition in order to provide a suitable surface for your solar project.

Table 4.6 provides general guidelines for how to assess your roofs based on their age and condition. Before proceeding with a project, have the roof inspected and review the report with the installer, roof maintenance provider, and roofing manufacturer. Get their feedback on installing solar and their commitment in writing to honor existing warranties after solar is in place.

While preferable, it is not critical to have a brand new roof on your building for it to be suitable for solar. Many roofs are highly durable and can be expected to last for the life of the solar array. Roof coatings can be applied to some roofs in order to extend their lives. In other cases, it may be possible to plan for the solar array to be moved in the event that extensive roof work becomes necessary. This cost will need to be underwritten with the project. For these reasons, it is a good idea to start out with a new roof under the solar array whenever it is feasible.

Roof type

The type of roofing on your building influences the design, and therefore the cost of your solar project. The most common types of commercial roofs are:

- membranes (examples include: EPDM, TPO, PVC);
- built-up modified bitumen;
- preformed or standing seam metal;
- monolithic (examples include spray-applied, fluid-applied).

Table 4.6 Roof assessment guidelines

Roof age (years)	Remaining life (years)	Suitability for solar	Guidance
Less than 5	15–20	High	Perform preventative maintenance prior to solar installation
5–10	10–15	Low to moderate	Consider: roof coatings to extend roof life; delaying solar project
10–15	5–10	Low to moderate	Consider: accelerated roof replacement; delaying solar project
15–20	0–5	High	Accelerate roof replacement schedule to coincide with solar project

Almost any roof type can support solar, although buildings with less durable roof membranes can present an increased risk of future problems. This is of particular importance when you consider that you may be covering the majority of the roof with solar equipment that is not easy to move in the event a leak does occur. Roofs that have deferred maintenance should be reviewed carefully before adding solar, especially if the roof is currently having problems with leaks. Regardless of whether a roof is new or not, ongoing and future maintenance needs should be considered in light of the additional equipment being added to the roof.

White reflective roofing is a good companion to solar projects because white roofs do not get as warm as dark-colored roofs such as modified bitumen or some membrane roofing. This matters because solar modules produce more electricity if they stay cool. A dark-colored roof could reach a surface temperature of 150° or more on a sunny day. In the same conditions a white roof may not even reach 100°. Elevated roof temperatures not only affect the amount of energy produced by the solar array, they also cause greater expansion and contraction stress within the solar equipment and the roof itself. This can lead to differential movement and abrasion where solar equipment contacts roof surfaces. In extreme cases problems can even arise with the solar components themselves due to excessive temperature-induced movement in conduits, cable trays and other equipment.

Roof warranty

Many people working in the solar industry place a great importance on roof warranties. A common approach you may see is to declare any reasonably new roof with a 15– or 20–year warranty a good solar candidate. This is an over-simplification that can be

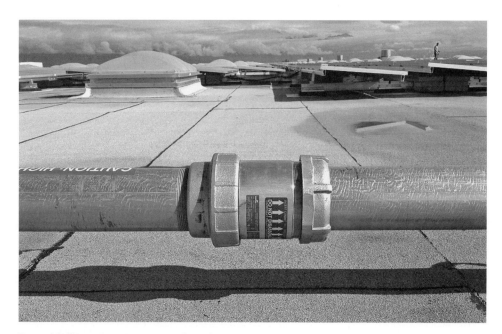

Figure 4.1 Expansion joint in a metal conduit.

Lessons from the field

Protecting ballasted membrane roofs

You may find that ballasted membrane roofs require additional care when considering solar. These roofs have a single ply membrane covered with river rock or other small stones that help protect the membrane from damage due to sun, wind, and weather. When installing solar, the added foot traffic can cause the stones to begin to break up. Stepping on these broken stones can lead to roof punctures and eventual leaks. This can be managed effectively by removing ballast and cleaning the roof before commencing work, but this step increases project costs. Gravel-ballasted modified bitumen roofs tend to have fewer problems due to greater membrane thickness and a self-healing ability inherent in the asphaltic roof materials.

potentially risky for the property owner because it shortcuts the necessary physical review of the property, identified in this chapter. It also tends to de-emphasize the potential costs of maintaining the roof in the future once solar is installed. Document both the roof warranty's term, as well as the expected remaining roof life. Note that these may not always coincide. A poorly maintained roof with a 20-year warranty may need replacement after 15 years, while a 15-year roof that is well maintained in a moderate climate may last for 20 years or more. Include both values in your portfolio due diligence spreadsheet.

As the previous example illustrates, a roof warranty may be of limited value when considering solar because it does not reflect the actual roof condition based on maintenance, climate, and use. The fine print of most warranties also contains exclusions for excessive foot traffic, incompatible materials that touch the roof, and modifications to the roof, such as placing equipment atop the roof membrane.

As a roof nears the end of its warranty, it also tends to become increasingly difficult to get service for the warranty from manufacturers who would prefer to sell you a new roof, and from contractors who have long since moved on to other projects. That said, addressing roof quality and suitability should not be minimized. It is a critical risk mitigation step for property owners. There are ways to manage these risks that will be discussed in Part IV. While roof warranties are included in the property assessment, they are a secondary factor when compared to real-world roof quality and expected longevity as determined by a detailed roof inspection.

Roof height

The height of a commercial building roof can affect the cost and planning necessary to install a solar project. Most or all of the solar array components will need to be delivered to the roof by some type of lift or a crane. For buildings with roof heights greater than about 30 feet (ten meters), a crane will be needed. The designer of the solar array on a tall building may also be required to account for higher wind forces acting on the solar modules. The height of a building is unlikely to jeopardize the solar project's viability, but it can increase costs and dictate certain choices in the design and construction of the array.

Rooftop equipment

For the portfolio due diligence review, note the amount of equipment on the roof. In the solar industry, equipment or other objects that impact the ability to site solar modules are referred to as "obstructions" on the building's roof. These rooftop obstructions are anything that takes up space on the roof – skylights, heating and cooling equipment, billboards, vent stacks, elevator penthouses, and the like. For your initial assessment it is sufficient to note the level of obstructions very generally, such as "minimal," "moderate," or "extensive." Roofs that have minimal equipment tend to be better candidates. Moderate rooftops may be fine, while extensively cluttered rooftops can present challenges when it comes to finding enough space for modules, as well as for installing them in an expedient manner. This can drive up costs and make maintenance of the solar array and the roof itself more difficult. If you do not know how to classify the roof, a quick review of the building on an aerial satellite image application like Google Earth will give you a good idea of how much rooftop equipment exists on the building.

Figure 4.2 Rooftop obstruction.

Note: the solar array has been designed to accommodate a roof with extensive skylight coverage by providing adequate setback space on all sides of the skylights.

Tenants

The contractual relationships between the landlord and building tenants are essential to consider when looking at solar projects. I have yet to find a landlord who is willing to lose a building tenant as a trade-off for installing a solar array. This is because the value of even a small tenant can be comparable to the value created by a solar array. Recall the due diligence considerations that you included in your due diligence spreadsheet:

- tenant name and contact information;
- lease expiration date;
- lease type, such as triple-net, full service gross, etc.;
- party responsible for roof maintenance, either owner or tenant.

For the due diligence assessment, note the tenants and their lease expirations. This will be particularly important in cases where you plan to sell solar electricity to building occupants. Even if you are not interested in selling power to tenants, knowing lease expirations is useful to manage their expectations and to make construction easier. It also enables you to plan more effectively in case you need to modify lease language to support your solar plans, or if you would prefer to have the building vacant at the time of the solar project. For instance, you may have a large tenant that is leaving in the next year. Rather than try to work with the tenant to negotiate an amendment to their lease or to coordinate the installation of a solar project around their use of the property, you may decide to wait until they leave to begin the solar project.

Include data in the due diligence review on the type of leasing arrangement you have. Is it a triple-net lease where the tenants are responsible for their own energy bills? Is it a gross lease where you provide energy to the tenants? This will matter when it comes to deciding what you want to do with the electricity produced by your solar array. It may be straightforward to use solar to offset energy needs in a full service gross-leased property, but daunting to make that arrangement work in a triple-net property that has many tenants. Part III describes the project structures available to suit these situations.

Lessons from the field

The silver lining of tenant vacancy

In the case of solar, a tenant vacancy offers flexibility that can work in your favor when it comes time to install the solar array. Vacant and unused parking spaces can serve as staging areas and parking for construction vehicles. There is no need to re-negotiate existing leases. One less tenant means less potential for complaints about noise or other disruptions. Installing new electrical closets will be easier if they are in or near a tenant area that is unoccupied. Depending on the property type and location, you may even be able to use the vacant space to stage solar construction materials.

Tenants' rights to the area of the building where you want to put solar can affect project feasibility. Tenants, particularly in single-tenant buildings, may have rights to use the roof or parking area – the two places where solar is likely to be planned. Landlords seeking to use spaces for solar projects that generate energy for their own needs will have to first address the tenants' rights to those leased spaces. This is less of a problem on multi-tenant buildings where the landlord is in charge of maintaining the roof and other areas of the site. But even here, multi-tenant leases may allocate roof maintenance obligations based on their pro rata share of space. Adding solar to a roof that tenants are charged maintenance fees for could be seen as an added cost burden that they are unwilling to pay. This is particularly true if the tenants are not receiving the electricity produced by the solar array. In this situation, it only takes one tenant to say no to jeopardize the solar project. Even if the tenants are supportive, you will still have to work with them to maintain support throughout the planning and construction process.

For solar projects where you intend to sell energy to tenants in the building it is important that the remaining term of the tenants' leases is considered. When it comes to selling power to tenants, in most cases the longer the lease the easier time you will have arranging the solar project financing. This is important because many solar projects use third-party financing. The revenue to pay back this financing often comes from the sale of power to tenants, so the quality of this revenue is very important. This is referred to in the solar industry as "power sales risk." There are two primary factors that contribute to power sales risk.

The first factor is the term of the power sales agreement. Projects typically seek to align the term of the power sales contract with their solar financing. Financing often requires a 15 or 20 year term, so the power sales contract should seek to align with that. A power sales contract shorter than the term of the financing introduces risks for the financial partner that few are comfortable taking. A triple-A credit-rated energy off-taker that only has a ten-year lease is unlikely to attract a great deal of interest in your project from providers of solar financing. That same customer with a 20-year lease would be an ideal financing candidate.

The second factor is the credit quality of the counterparty to the power sales contract. The higher the credit quality of the energy off-taker, the easier it will be to find third-party financing for the solar project. Many solar financing providers are unwilling to provide financing if the off-taker is not an investment-grade credit risk. This means that off-takers that are small businesses, professional service firms, or companies with little or no credit rating may make it difficult to obtain solar financing, even if their business is stable and profitable. While some financing providers offer a degree of flexibility in this area, they would invariably prefer to finance a project with a strong credit-rated power purchaser.

The combination of the two elements – lease term and credit risk – can be looked at together to identify good candidate locations for third-party solar financing. Table 4.7 provides a matrix that illustrates the interaction of the two factors. You can expect that many properties will not fit the mold of a project where third-party finance is likely to be a good fit for solar.

This is also likely to be the case for projects where the building owner is the energy off-taker. Buildings are often controlled by single-purpose entities such as limited partnerships or limited liability corporations. These entities are not credit-worthy on their own, so the financing provider has to look through the entity to the parent company in

Table 4.7 Solar financing feasibility

		Credit rating		
		None	Low	High
Lease term	Short	Poor	Poor	Poor
	Medium	Poor	Poor	Moderate
	Long	Poor	Moderate	Good

order to find sufficient security to agree to provide financing. This may not be feasible for small, privately owned property portfolios, where even the entire business has little or no credit rating. Another issue that can arise is that providing that look-through security for the solar financier means that the single purpose entity is no longer a stand-alone when it comes to the solar project. For these reasons it can be difficult for many property owners that want to deploy solar with third-party financing to do so successfully.

Bundling several tenants in a multi-tenant building is sometimes looked at as a way to overcome the lease term limitation and off-taker credit risk. The logic is that diversifying across multiple off-takers provides some certainty that there will be at least a few tenants able to use solar energy. This may work in theory, but it is extremely difficult to pull off in practice. This is because there is a much greater burden on the financing provider to underwrite numerous off-takers, overlay their lease terms, and then get comfortable that they will all be able and willing to purchase power from the solar array. Adding this all up is sufficient to lead most financing providers to seek simpler projects elsewhere. Power sales risk tends to be difficult to overcome in multi-tenant buildings unless the energy demand is significantly greater than the system output. To address this limitation there are a growing number of solar project structures that do not require tenants to purchase energy in order for a rooftop solar project to be viable.

Energy use

Solar arrays produce electricity throughout the year and you need to have a plan for what to do with the energy. Some building owners use it for their own needs. Others sell the electricity to tenants, either directly by billing them as a second utility provider, or indirectly through common area maintenance charges. Still other property owners sell electricity as a commodity directly to the electric utility under a long-term contract, bypassing the building and its tenants entirely. You do not need to have your approach figured out at this point, but you will want to collect basic information about the building's energy use in order to evaluate the options.

If you are considering using the energy from the solar array to offset the energy used at your building, you will need to find out the annual electricity use of your property. This is the total kilowatt-hours of electricity the building consumed, for all areas of the building that you want to serve with your solar array. This will often include more than one meter or sub-meter. This may also include more than one utility account. For example, you may want to offset common area electricity use. There could be one account for outdoor lighting circuits, another account for indoor common area electricity needs, and others for central plant equipment and back-of-house spaces.

Gathering the data to calculate whole building electricity use can present challenges. If you have building tenants that pay their own energy bills, you will need to get copies of their electric utility bills, typically for the past year at a minimum. This request can take days or weeks, depending on how eager the tenants are to help out, how accessible their records are, and how many meters they have. If you happen to control the energy service for the property in question, find out if you have a building energy management system that monitors building energy performance. This system may have the data you are looking for.

If you manage your buildings but only pay for common area energy use, as with many triple-net leased properties, you may find that your accounting system is set up to manage the cost of energy, not the amount of energy consumed. Records of utility payments may be in your accounting system, but there may be no data on how many kilowatt-hours were consumed. In this case, you can either get the property manager to provide hard copies of the utility bills, or you can contact the utility provider directly, give them the account number, and request the energy history for that meter over the past year or two. As long as the meter has been in the same name for that amount of time, they should be able to provide the data for you.

One resource you may be able to look to for properties in the United States is the Energy Star Portfolio Manager tool. If the tenant or a property manager has set up an account for the building in question, you can retrieve the current energy use data of your property from the online Portfolio Manager dashboard once they provide you with access. The Portfolio Manager can be accessed at: www.energystar.gov/index.cfm?c= evaluate_performance.bus_portfoliomanager

Electricity price

The price of electricity can also affect project selection between one market and another, so it is important to include electricity prices in your due diligence spreadsheet. This is particularly important if you think you will want to offset your own energy needs or sell electricity to building tenants. When electricity is sold to building tenants the price charged is often equal to or at a discount from retail energy rates. This rate can have a significant impact on project revenue, although it is unlikely to invalidate a project entirely. For gross leased buildings, landlords have visibility into their energy costs and the rates they charge their tenants. For triple-net building leases, where tenants may pay different rates based on their size and electricity usage, it can be more challenging to establish the value of electricity produced by the solar facility.

Price data can be found on utility bills. There may be a base rate for the electricity composed of various production and transmission charges, plus other fees, taxes and surcharges. If these are used to calculate the total cost of each kilowatt-hour of energy consumed, include them in your calculation of the unit cost of electricity. Fixed charges do not need to be included since they have to be paid regardless of the amount of energy consumed.

Knowing energy prices is also important in the event that there is excess energy generated that can be fed back onto the grid. In the United States excess energy fed back to the grid may not be purchased by the utility provider, but may instead be credited against future energy consumption. If it can be sold, it is often priced at a wholesale rate as much as 50 percent or more below retail electricity prices. These

Lessons from the field

Electricity prices in the United States

Electricity prices in the United States vary by more than 500 percent by state from the lowest price to the highest price. Table 4.8 lists the range of electricity prices. Current data on electricity prices can be found online at the Energy Information Association website at: www.eia.gov/electricity/sales_revenue_price

Table 4.8 Electricity prices by state

State	Electricity price (US cents per kilowatt-hour)	State	Electricity price (US cents per kilowatt-hour)
Hawaii	32.4	Montana	9.1
New York	15.8	New Mexico	9.1
Connecticut	15.6	Nevada	9.1
Alaska	15.1	Texas	8.8
Massachusetts	14.3	Kansas	8.8
New Hampshire	14.0	Indiana	8.8
Vermont	14.0	Illinois	8.6
New Jersey	13.5	Minnesota	8.6
California	13.1	Kentucky	8.5
District of Columbia	12.9	Louisiana	8.4
Rhode Island	12.4	Oregon	8.2
Maine	12.3	West Virginia	8.1
Maryland	11.3	North Carolina	8.1
Delaware	10.6	Missouri	8.0
Alabama	10.5	Nebraska	8.0
Wisconsin	10.4	Virginia	8.0
Michigan	10.3	Iowa	7.9
Tennessee	10.3	South Dakota	7.8
Pennsylvania	10.0	Wyoming	7.7
Georgia	9.9	North Dakota	7.6
Florida	9.9	Oklahoma	7.6
Ohio	9.6	Arkansas	7.5
Arizona	9.5	Washington	7.5
Mississippi	9.5	Utah	7.4
Colorado	9.4	Idaho	6.4
South Carolina	9.3	US average	10.2

rules create financial disincentives to operate solar facilities any larger than needed for the building.

Final thoughts

At this point you will have gathered the information necessary to complete your property due diligence assessment. You should have already identified any properties that are unsuitable for solar due to operational or asset management considerations. Other physical building, tenant, and energy factors have been identified and are likely to provide you with an intuitive sense about which properties are likely to be more suitable candidates than others. For example, a property with a small roof area and a roof system that is in the middle of its life is likely to be less suitable than one with a large roof that was recently replaced. A large, high-energy-using tenant may be more interested in solar than a tenant that uses very little electricity. Your intuition will help prioritize feasibility for your portfolio, but you will also need additional perspective from those in the solar industry to fill in the blanks on incentive programs, project costs, and other factors that complete the process of prioritizing your property list.

The most efficient way to find the solar information you need is to reach out to solar industry professionals with your portfolio due diligence spreadsheet in hand. The resources and knowledge of the solar industry will allow you to begin focusing on the properties and projects you want to pursue. Don't worry if the spreadsheet is not entirely complete – the information-gathering work up to this point will greatly assist solar service providers in providing you with real-world solutions that work with the portfolio you have, not a solution that relies on generic assumptions for that market.

So far none of the property due diligence discussed in this chapter has required you to hire consultants or incur any significant out-of-pocket costs. The structural analysis and other inspection costs can be held until later in the project. At that time you will have a solar service provider involved that will be able to spearhead the structural analysis and coordinate other due diligence. With the portfolio assessment in hand you are ready to put the data to use and begin reaching out to discuss your portfolio with service providers, solar developers, and other resources in the solar world. Their input will allow you to better define your options for deploying solar within your portfolio. Chapter 5 discusses what to look for in these companies.

Chapter 5

Solar service providers

Chapter summary

- Solar service providers are companies that help you plan and execute solar projects. The support they provide can range from a single professional service to full turnkey project development and financing solutions.
- Select a solar service provider that has demonstrated expertise in the markets you are interested in and offers the scope of services you seek.
- Identify solar service providers through your professional network such as colleagues, sustainability managers, attorneys, real estate service providers, financial institutions, and local utilities.
- Review the health of the solar service provider, including financial resources, partners, and operational track record for prior projects. Ask for references.
- Verify the assumptions made in financial models provided by solar companies. Ask for historic data or comparables if you have doubts.
- Find out who the long-term owner of the solar array is expected to be. Unless you purchase the solar array, expect the solar developer to sell the project to a third-party owner upon completion.

You are now ready to reach out to companies in the solar industry that can help finalize the selection of sites and take your project from vision to reality. These are the solar development companies, engineering consultants, solar financiers, contractors, vendors, and others with expertise in the solar space. Collectively, they are referred to throughout this book as *solar service providers* or *solar developers*. With the portfolio due diligence process complete, you are equipped to have focused and productive conversations with these service providers. They will be able to help fill in the blanks by giving you the lay of the land in terms of incentives, financing, project ownership structure, and technology. Combining their solar sector knowledge with the knowledge you have of your portfolio will allow you to identify the properties where it makes the most sense to pursue solar projects. Because of the importance of the support they can provide, selecting these solar service providers is a critical step in producing a successful solar project. This chapter walks you through the key selection criteria for finding the best solar service provider for your needs.

Service provider capabilities

Your portfolio assessment likely identified some properties that appear to be an initial fit for solar projects. Now you are in a position to seek out the expertise that a qualified solar service provider can offer. Look for solar service providers that allow you to augment your knowledge of your property portfolio with their experience in the solar sector. To identify the companies that are best suited to help you execute your project, seek out companies with some or all of the following characteristics:

- established presence in the markets where you have interest;
- experience working with the commercial real estate industry;
- in-depth knowledge of solar incentive programs;
- strong project management experience;
- solar array design and engineering capabilities;
- diversified procurement channels;
- pre-established project finance capability;
- ability to provide operations and maintenance (O&M) services.

For many solar projects on commercial buildings, working with an experienced solar development company is a practical solution for project delivery. Their solar-focused experience managing a complex and specialized design, permitting, and construction process work in your favor. These companies often specialize in providing engineering, procurement, and construction (EPC) services. These services are typically a turnkey package that bundles the majority of the expenses you will incur to develop a project. The solar developer becomes the responsible party for integrating the required elements necessary to deliver a successful project on your behalf.

Look for solar developers with experience on projects of a similar scope and size that have been completed in your market within the same utility territory. Ask to speak to references to get more familiar with their track record. Solar project owners often enjoy sharing their experiences, both good and bad, so this can be an easy way to gain a new perspective on the firm and the project you are about to undertake. Ask to visit a completed project and talk with the property owner's project manager so you can hear from them directly about how the project turned out. This will also give you an opportunity to envision what a completed project could look like on your property. Each company will have different strengths and experiences. It is wise to find a company whose expertise augments and complements your own knowledge.

You may wonder if you really need to talk to a solar company when you have capital and internal staff resources to figure out solar projects on your own. While you may eventually not need a solar service provider on every project, the experience and industry relationships a good partner brings to the table are difficult to replicate. This is particularly true when you are pursuing your first project. An experienced solar service provider will bring knowledge to the table from scores of project. This track record is essential during your first projects and will significantly reduce risks and costs. This is especially important in the fast-moving solar industry of today.

Capabilities

Solar service providers come in all shapes and sizes. At one end of the spectrum, you may find a local contractor who has teamed up with an investor, working together to execute projects on an ad hoc basis while they also do their day jobs elsewhere. You'll also find system designers, engineers, and financial advisors that specialize in solar. Toward the middle of the spectrum are regionally focused solar EPC companies that service both residential and commercial projects with an in-house staff and established relationships with solar product suppliers. They will often have standing financing agreements with third-party investors and banks. In this same range are private investment groups that have established relationships with vendors and installers to service an existing real estate client base looking for solar from existing financing partners. At the other end of the spectrum are full-service global solar development companies that provide turnkey solar solutions that include planning and design, procurement, and financing via their own in-house resources. Some of these companies are subsidiaries of solar module manufacturers. This gives them access to deep product expertise, global resources, and strong financial backing compared to many smaller companies. These companies offer turnkey EPC services for large commercial and institutional projects while using wholesale distribution channels to reach smaller projects in the commercial and residential sectors.

Up and down the broad range described in this chapter you will also find real estate service providers and engineering firms with solar engineering divisions that can support some or all of your project needs.

It may seem like a challenge to figure out what type of service providers offer the right resources and expertise for your project. The basic types of solar companies can fortunately be categorized to aid your search. The categories in Table 5.1 are not a formal industry-recognized classification system, nor do they imply that one is better than another, but they do provide some structure to an otherwise confusing group of industry participants.

Table 5.1 Solar service provider groups

	First tier	*Second tier*	*Third tier*
Markets	Local or regional	Regional or national	Global
Services	Limited to their specialization; others sub-contracted	Core services provided in-house; limited to moderate use of sub-contracting	All major services provided in-house; selective use of sub-contracting
Financing	None, or project-by-project; usually provided by a third party	Project-by-project or aggregated; pre-established financing solutions	Aggregated; pre-established financing solutions; may use in-house capital
Project scale	Small to mid-size	Small to large size; plus limited utility scale	Mid-size to large scale; plus frequent utility-scale
Service integration	Minimal vertical integration; high degree of specialization	Vertically integrated service business (sales, engineering, procurement, construction, finance)	Full vertical integration (distribution channels, sales, engineering, procurement, construction, finance)

First tier

First-tier service providers are characterized by a focus on offering a specific expertise or service within the solar industry value chain. There is no rigid definition of what this includes, but these service providers tend to focus on specific markets where they have deep knowledge of local programs and market conditions. Because of this, they tend to be better suited to provide services in a few local or regional markets. First-tier companies typically have expertise in one or a few areas, and they outsource other services in order to complete projects. An example is an electrical contractor that offers solar installation and maintenance because it overlaps with their electrical installation expertise, but outsources project finance and engineering. In another example, a private investment group may provide an efficient capital structure, but outsource the design and construction of the solar array. First-tier companies tend to be smaller shops and focus on projects of a few kilowatts up to perhaps 250 kilowatts. That said, their financial and staffing resources could be stretched trying to execute large projects if they are a small company. These companies are likely to have a preference for products that they have experience working with, although they can use products from comparable manufacturers wherever prices are most competitive. Large projects and those with a great deal of complexity in the design or execution could present challenges for first-tier providers.

Second tier

Second-tier service providers are characterized by a regional or national focus. They generally offer full EPC services in-house. They have pre-established financing capabilities from third-party capital sources. This gives them the ability to execute a broad range of projects, from small arrays of a few kilowatts up to a few megawatts. Second-tier providers often have deep regional market expertise and can offer hands-on management that aligns well with mid-sized projects on commercial buildings. These service providers can be valuable in comparing incentive programs across your property portfolio if they also operate in the applicable markets. These service providers tend to have a few preferred technology providers that they have had success with and have therefore established relations with to provide reliable supply and discounted pricing. These service providers are likely to have pre-established financing arrangements with third-party capital sources. Second-tier service providers tend to be small enough that they will be interested in almost any project you have in mind, and large enough to offer a range of EPC and financing solutions.

Third tier

Third-tier service providers tend to be the largest in the business and offer a full range of services on a global basis. They have market knowledge that allows you to compare solar opportunities across national and global markets. This can be particularly valuable if your portfolio spans more than one country. They have in-house teams that focus on large commercial projects, as well as other groups that are focused on much larger utility-scale solar projects. They are likely to have established in-house project financing that they use for their own utility-scale projects, as well as for large projects such as those on commercial buildings. Their pipeline can be many times larger than first- and second-tier companies due in large part to the utility-scale projects which could be hundreds of megawatts. This can lead to beneficial economies of scale when procuring equipment.

Third-tier service providers are unlikely to target projects smaller than perhaps 100 kW unless there are several smaller projects that can be bundled together. These providers are the most vertically integrated and may even have access to upstream solar product manufacturing within their parent company. Due to the vertical integration with a manufacturer these companies may be affiliated with a specific product. For

Lessons from the field

Doing a financial check-up on your service provider

The solar industry is growing quickly, and that means a lot of solar service providers that are here today may not be here tomorrow. An untimely bankruptcy or an unexpected handoff from one service provider to another can reset your project's schedule and reshape the project team you are used to working with. At worst, it can leave you holding a project with no one to complete it.

Your due diligence for solar service providers should not only focus on their expertise and professional capabilities, but also their balance sheet, how long they've been in business, and the health of any parent company or financial backers. A few questions to look into are:

- Are they sufficiently well funded?
- Is their business growing?
- Are financial partners and vendors stable and well capitalized?
- Are they large enough that if they do fail, another company will want to take over their projects?

When you consider these elements, you will need to decide which tier of service providers makes you feel sufficiently comfortable that you are not taking on too much risk in terms of whether the company you are working with will be around to see the project through. In the solar industry today it is difficult to eliminate these risks completely, but there are ways to minimize these risks. The most effective ways include:

- working with the large, well-established service providers;
- diversifying providers and using more than one to meet your solar needs; perhaps by having a different provider in each market;
- compartmentalizing the services required for the project. That way, if one service provider cannot perform, you could replace them without jeopardizing other aspects of the project.

projects that have a significant regulatory compliance requirement, such as those selling power directly to a utility, third-tier companies generally have strong expertise dealing with utility regulations and documentation requirements.

There are other service providers that fill in the gaps and provide services tailored to the real estate industry. National and global engineering firms and real estate service provider companies have solar service groups that provide fee-based consulting and advisory on solar projects. Their services range from basic market comparisons and consulting services up to full turnkey project execution on behalf of the property owner.

Categorizing solar service providers in this way enables you to better understand the expertise each one can bring to your project. With this information you can identify the capabilities that are most important for your project needs. No one tier is intrinsically better or worse than another. For example, you may have a geographically concentrated portfolio, you are focused on a few smaller projects, and you want to use your own consulting relationships. In this case you may only need a first-tier service provider to perform a limited scope of work. At the other end of the spectrum, you may have a global portfolio and you want to evaluate solar opportunities across countries, not just between individual buildings or markets. Working with a third-tier service provider will likely be more productive than spending your time and effort identifying good solar service providers in each market and then aggregating their recommendations.

Finding solar service providers

If you own properties in a market that has a healthy solar incentive program, the chances are good that you or someone in your organization has already been contacted by a number of solar service providers. These solicitations are one easy way to find a company to work with, but there are many other, better ways to find a service provider suited to your needs. While they require more effort than just waiting to see who walks through the door, you are much more likely to come away with a well-matched service provider relationship.

- Use your *professional network* in the real estate industry to identify service providers. Talk to other property owners, investors, developers, and asset managers. Anyone in this group that has looked into solar is likely to have a perspective similar to yours, and can be a valuable resource. Ask what service providers they have talked to and what their impressions were. Do the same with architects, engineers, general contractors, electricians, roofers, and other vendor relationships you have. Inquire with real estate brokers, but be aware that they may steer you to companies they have a financial relationship with. References from your professional network serve as a good starting point. It does require care to ensure you find the right fit rather than just the easiest introduction.
- Talk to *people involved in sustainability* for real estate. This can be a person within your organization, or you can ask for a referral through industry colleagues. Many real estate industry associations have sustainability committees and resources. Committee members can point you to sustainability experts in the industry. Across the board, you will find that sustainability professionals will be more than happy to share their knowledge and solar industry contacts with you.

- *Real estate service providers* can be a good resource for solar services. Global service firms such as CB Richard Ellis and Jones Lang LaSalle, among many others, have developed in-house service offerings in response to the demand for solar products and services. They provide one-stop shopping that can greatly simplify the process of planning a solar project. They can also provide other defined scopes of service based on your needs, such as market analysis and relevant industry research.
- Your *contacts at financial institutions* are another good source to identify solar service providers. If the financial institution has done business with solar companies they will have vetted the company and the sector. This is particularly useful if you are looking for second- or third-tier service providers that are likely to have had significant involvement with the financial industry. These contacts will be in a good position to steer you toward companies that they are comfortable lending to and financing projects for. Some financial institutions, particularly larger banks, have entire divisions dedicated to financing solar projects. Investment banks such as JP Morgan, Goldman Sachs, and others have investment funds that invest extensively in solar projects sourced from leading solar project development companies that you may also be considering.
- Reach out to *contacts at law firms*. Attorneys are at the forefront of negotiating agreements between property owners and solar companies. The firms you work with may be able to point you to companies they've dealt with successfully. If they are at a large firm they are likely to have someone in-house that has worked on solar transactions for several property owners. If not, they may be able to direct you to an industry colleague who has worked on solar contracts.
- *Electric utilities* can steer you to additional resources for solar programs and also solar service providers. While the utility is unlikely to recommend any particular service provider, they may be able to tell you where you can find out what companies have been actively submitting solar applications to the utility's solar incentive programs. Begin by contacting your utility account representative and ask them to refer you to the person in charge of solar energy programs for commercial buildings.

Online resources for finding service providers

Another way to find solar service providers is to scan the press releases that publicize completed solar projects. This allows you to identify companies that are active in the market where your property is located, and it allows you to focus on those working on projects like yours – on commercial buildings of a particular size. A press release will typically identify the solar developer as well as the location and the size of the solar project. It may even identify the property owner, financier, or other contributors to the project. Contacting these references is a useful way to find out more about the capabilities and performance of that particular service provider. There are many online resources where you can find press releases. A few of the most widely read are listed below:

- www.solarbuzz.com
- www.solarfeeds.com
- www.solarplaza.com
- www.solarindustrymag.com

- www.pv-tech.org
- www.electroiq.com/index/photovoltaics.html

To expedite your review of press releases after you have them onscreen, you can use the search function of your web browser to locate keywords in the press releases. Keywords include the market where your property is located and the type of project. For example, you could search press releases on solarfeeds.com for "Massachusetts" if you own properties there. You could also search for "roof lease" or "power purchase agreement" as a way to begin to sort through project announcements for companies working on projects similar to what you are looking for. The downside of this approach is that not every solar company issues a press release when they complete a project, so you have no guarantee that your searches are providing you with a full list of qualified solar service providers.

You can also search for solar service providers through the solar industry associations of which service providers are members. The Solar Energy Industries Association (SEIA) has a member database online that has searchable features. The SEIA website can be found at: www.seia.org. Search the member directory for "project developers" as a starting point. The Solar Electric Power Association (SEPA) also has resources for the solar industry, although its member directory is only available for a fee: The SEPA website can be found at: www.solarelectricpower.org. These organizations are United States-focused, but similar organizations exist in most countries with active solar programs. They can be found by searching the web for "solar industry association" or "photovoltaic industry association" for your respective country.

Quick tip

Industry association directories speak little to the qualifications of the service provider other than that they paid a fee to become a member. But even this low hurdle does at least indicate that they have gone to the trouble to invest in joining their industry association.

You can also perform searches online for solar companies in your market, but these may take a lot of effort before they provide much real value. You can, however, use the internet to gain additional information about solar companies that you have identified through other means. While it may seem obvious, companies with a poorly designed website that is missing comprehensive information about the company, their capabilities, and their project experience should raise a flag about their resources and experience. While this is not a given, it can be informative when considered along with other factors.

By using a combination of these sources identified here you will be able to identify a number of solar companies that may be a fit for your project. You may find it useful to categorize them into the three tiers previously discussed so you can organize your search. You are likely to find that this process of comparing them side-by-side will

begin to give you an intuitive sense about which service providers have the right fit
for your needs. Knowing this will save you time, ensuring that you spend less time
talking to companies that are not well aligned with your goals. Once you have a sense
of the companies that seem to be an initial fit for your project, you can reach out to
them to continue the due diligence process.

Your first conversation with a service provider

Call or email the service provider. Tell them that you are looking for a company that
can assist you in vetting the opportunity for solar projects for your property portfolio.
When you speak with them, keep the following questions in mind:

* What markets are you actively working in today?
* What services does your company specialize in providing?
* What services are outsourced?
* What is the typical size range of projects you've completed?
* Have you completed projects on commercial buildings? How many and where?
* Who were the property owners and can you provide referrals?
* How long have you been in business?
* Is the company affiliated with a non-solar business, such as electrical contracting,
  roofing, private capital, or other business?
* Do you provide financing solutions for solar projects?
* What company provides the capital for the financing?

The service providers will want to know what your objectives are for solar projects.
Are you looking to reduce your own energy use, sell power to tenants, or generate revenue
by selling electricity to the utility? They may ask if you are looking to purchase the
solar array or if you want to finance it – either through a power purchase agreement
(PPA) or a solar lease.

Based on the initial calls you will begin to develop a sense of which service providers
seem like a better fit for your needs. Create a shortlist of companies you are interested
in talking to further. Set up a follow-up call or meeting with them to discuss the specific
opportunities within your portfolio. At this stage you will want to provide your
portfolio due diligence spreadsheet to the shortlisted companies so they can provide a
high-level analysis of the opportunities they see in your portfolio. The detail you have
compiled in the portfolio assessment will make this process much more time-efficient
for them, reduce the number of questions they will come back and ask you, and more
quickly get you to a useful list of target opportunities.

Quick tip

When you send your portfolio due diligence spreadsheet to a service provider you
may want to edit out any confidential information. Alternatively, you can ask them
to sign a non-disclosure agreement.

Your first meeting with a service provider

When you meet with the solar companies, you will want to gauge how their market knowledge and prior experience matches up with the projects you have in mind. The list of questions we provided earlier in this chapter provides a good starting point for evaluating company qualifications. Your objective is not only to see if they meet an objective standard of capabilities and expertise, but also how they approach projects. Are they methodical and conservative in their approach to underwriting, engineering, and construction? Or do they build aggressive assumptions into their financial models and schedules? Do they have an established financing solution or will they create one specifically for your project? Do they push aggressively to win your business or do they focus on educating you about the opportunity? These questions don't have universally right or wrong answers. Depending on what you are looking for, the various approaches will either resonate or not. Your own experience and intuition will serve as a good guide for identifying a company that is compatible with your own attitudes and goals. The next section reviews topics that will give you a clearer picture of how your shortlisted companies stack up against each other.

Key due diligence considerations with service providers

Solar service providers can provide you with a project financial model showing the project's costs and expected revenues over its lifetime. Review the numbers as you would

---

Lessons from the field

Investment requests

Over the years, more than a few companies have contacted me with offers to develop solar projects – and to ask if my company would be willing to invest in their solar businesses or in other solar projects they were developing elsewhere. This is substantially more risky than simply owning or hosting a solar project.

As a variation on this, there are companies that ask you to co-invest in a partnership entity that owns the solar project that will be installed on your property. This is not buying the system or financing it, but taking a limited partner stake in the legal entity that owns the solar asset. This is typically portrayed as a way for you to gain in the upside revenue potential of a solar project and to align performance interests. In reality, it is often a way for the solar company to capitalize on the project while preserving its own balance sheet and maintaining control of the asset.

If you are not already a solar industry expert, ask yourself if it is wise to invest your capital into a solar energy business or in entities that own projects if you have little or no operational or management control. Fortunately, this type of request is becoming less and less common as the solar industry matures. There are plenty of high-quality and well-established solar service providers to choose from when the time comes.

a real estate development or acquisition opportunity. While this is not an all-encompassing list, ask yourself the following questions as you review the documents:

- Do growth rates seem reasonable?
- Do operating expenses look realistic?
- Will there be any major capital expenditures in the future?
- Is there a capital reserve?
- Are federal and state taxes on solar revenue accounted for?

These questions, as well as those raised in the following sections, will help you review the overall quality, accuracy, and applicability of the information provided to you by service providers. For a similar project, proposals should be fairly consistent overall, but it is unlikely that the assumptions from one service provider to the next will be exactly the same. As you evaluate them side-by-side, look for inconsistencies. Is one escalation rate much higher than the others, or is one maintenance budget lower than the average? These clues will help you identify areas where proposals warrant additional scrutiny in order to ensure you are getting the most accurate and realistic information available.

Electricity prices

Rosy assumptions in solar project financial models are fairly common. It is not difficult to find electricity prices projected to grow at 5 percent to as much as 8 percent each year for the next two decades. This is often used to show an ever-increasing value from replacing grid-supplied energy with the new supply of solar energy. While in a few markets – such as where deregulation recently occurred – electricity prices have jumped significantly, the long-term growth in energy prices is much less than some financial models would have you believe.

Take southern California, the poster child for high-energy costs and volatile prices. Even when including the unprecedented price spikes in 2001 due to the Enron energy price manipulation scandal, long-run electricity price inflation has averaged roughly 4.4 percent over the past two decades.[1] If the Enron-instigated market volatility were to be excluded, the long-run average would fall to around 3 percent.

Quick tip

It is easy to check energy price changes in your market.

1 If you have several years of energy data, review the electricity bills for your properties, remembering to look at long-term trends rather than short-term spikes due to deregulation or other one-off events.

2 You can download historic electricity price data for any area of the United States from the Bureau of Labor Statistics (BLS) website: www.bls.gov/cpi under the Consumer Price Index tables. These data have been tracked by the BLS for decades and will readily reveal long-term trends in your market. Look for "Electricity Price" data. Data for European countries is available at: http://epp.eurostat.ec.europa.eu/portal/page/portal/hicp/introduction

Capital expenditures

Check the solar financial model provided to you to ensure that it includes a realistic projection of capital expenditures. Some financial models show no capital expenditures whatsoever for 20 years. Even if it were possible, the solar project would be in such poor condition at the end of the 20 years that it would be severely underperforming and would probably be unsafe to operate. Inverters, key electrical devices that convert energy from the solar array to alternating current electricity, are typically warranted for 10–15 years. Realistic underwriting includes the cost for inverter replacement at some time during the life of the project, as well as a capital reserve to set aside money for the replacement. Depending on who is responsible for system operations, there could also be costs for replacing damaged panels, worn out wiring and the like. These costs are not likely to be included in the O&M budget and this type of expected wear is generally not covered by product warranties.

Tax incentives

Look carefully at the tax and other incentive assumptions that go into the financial model. Most solar developers make two fundamental assumptions that are rarely true to real life. First, they will assume that you are taxed at a corporate tax rate of as much as 40 percent, even though few commercial property owners in the United States are taxed at that high effective rate. The second assumption is that the project or you will be able to capture 100 percent of the tax and incentive value in the shortest possible time frame allowed by the government. Even if a third-party financing provider is engaged to monetize the tax benefits, the financier's inability to do so could affect the ability to obtain financing from the project.

Project schedule

Project schedules should be reviewed carefully. Is it realistic for a solar project schedule to go from initial meeting to operations within six months? Twelve months? Based on your experience getting entitlements and permits, negotiating contracts and leases, and working with engineers and contractors, you can decide if their schedule sounds realistic. In some places very short project schedules are possible; in the most active solar markets projects proceed at a fairly modest pace. Certainly, if you have a good working relationship, a template agreement, and you are simply duplicating it on subsequent projects you will save a considerable amount of time.

Quick tip

For your first project in the United States, plan on a 12–18-month schedule from concept to completion, and hope that you are pleasantly surprised if it happens sooner.

It is often helpful to have another pair of eyes look over the financial model. Ask a colleague that understands the development process or hire a solar consultant that has relevant financial modeling experience for similar projects. They can provide an impartial check on a financial model, tax assumptions, and other inputs in conjunction with your tax accountant, who can provide realistic assumptions for the particular situation.

Prior solar project experience

Solar project experience is two-fold; staff knowledge and the number of completed installations are good predictors of a solar company's ability to execute projects. There are few places to get a formal solar education. As a result, solar companies are often staffed with employees from non-solar industry backgrounds. Many engineers, contractors, real estate professionals, and investors have migrated to the solar industry as it has grown. As a result, solar companies may have skilled employees in certain areas but not others. They also may have a skilled team gathered together from other companies, but few completed projects of their own.

Look closely at the relevant expertise of the project team. Since many solar companies are relatively young, recognize the benefits and weaknesses of the person's prior professional career before they entered the solar business. A former banker may understand project finance, but do they have adequate construction and operational expertise to make you comfortable managing the installation of a large solar project? Professionals with electrical contracting construction experience can be your best allies during the installation of a solar array, but do they have the project finance know-how to put together a competitively priced and capital-efficient funding solution?

Look for the track record of completed projects executed by the solar company in the markets where you own properties. Few solar companies will turn down projects, but they may not have the expertise to handle the sometimes-complex regulatory requirements needed to capture incentives and comply with incentive program rules. Paperwork-intensive regulations and by-the-book utility rules can be complicated enough that even a small misstep can result in the loss of project incentives that can erode project profitability.

Lessons from the field

Financial health of solar service providers

I have seen the failure of more than a few solar companies over the years that I have been working on solar projects. The financial health of solar service providers is critical to the successful execution of your project. Take a close look at their finances. Distress or failure of their business may not directly impact the functional aspects of the solar equipment, but it could add complications down the road. Problems with previously completed projects are another sign that they could face additional financial stress as they are forced to rectify them. The importance of having a financially secure solar developer and operator is especially true when it comes time to sell the underlying property or refinance. Difficulties with the health of the solar project can create headaches for you and turn a value-creating asset into a potential liability.

It pays to talk to a number of potential solar service providers, ask direct questions, and vet their experience and capabilities for your project. Failure to do so could cost you money and leave you with a project that isn't meeting your expectations – or in a worst case scenario, one that doesn't get built at all.

Long-term solar project ownership

Solar developers often have arrangements to sell or transfer their ownership interest in a solar project to other investors seeking stabilized projects. These investors typically seek a long-term and low-risk source of income, so they prefer to take ownership only after the project is fully operational. This takes the asset off the solar company's balance sheet and allows them to recycle their capital for new projects. But unless you fund the project yourself, it means you will be dealing with a new owner that is not the solar company you worked with to develop the project. Unless you are planning to purchase and own the solar project yourself, ask the following questions:

- Where is the capital coming from to construct the project?
- Who will own the project after it is completed?
- At what point in the project's lifespan will this ownership transition occur?
- Who will provide ongoing operations and maintenance services?
- Are these standing contractual relationships already in place, or will they be created after the project is completed?

The goal of these questions is to find out whom you can expect to deal with as the solar project owner both today and in the future. You also want to find out what experience the long-term project owner has with solar. You want to be comfortable that contracts will be in place to ensure the project is maintained properly throughout its lifetime. If any of the answers cause you concern they should be discussed and clarified before proceeding with the project.

Final thoughts

Your intuition and real estate experience are good guides to choosing a solar service provider. If an offer sounds too good to be true, it just might be. Your goal is to find the solar company that fits your objectives and that has the skills and financial capabilities to deliver on their promises. When you are talking to solar companies, do not be afraid to ask questions, call references, or get a third party to verify assumptions. These efforts will go a long way toward ensuring you fully understand any risks you are taking and how best you can manage them to ensure the project meets your expectations. By combining your understanding of the capabilities of solar service providers in your market with the solar models discussed in Part III, you will be able to maximize the value provided by your chosen solar service provider.

Note

1   Calculation based on data from the US Bureau of Labor Statistics: http://www.bls.gov/cpi

Part II

Solar in real estate transactions

There are several points in the lifecycle of a solar project where its cost and the value it creates are magnified. These include:

- incorporating solar into a new development;
- adding solar to an existing building;
- property acquisitions and dispositions;
- leasing.

At these junctures, positioning solar properly helps to ensure that it creates increased value for all parties in the transaction without adding unnecessary complexity or risk to the deal.

Adding solar to a new development allows you to optimize the design of both the building and the solar array to capture the most cost efficiencies. Solar makes it easier to achieve sustainable building certifications. On an existing building, solar can be fine-tuned to match the operating profile of the property. This makes it easier to deploy projects with a high degree of certainty that the investment in solar will begin generating revenue quickly and reliably.

Acquiring or selling a building with solar is a key time to create value in the project's lifecycle. Acquiring a building that already has solar is an opportunity to capture the benefits of solar – clean energy, lower operating costs, or revenue. This comes without any of the risks associated with developing, financing, or otherwise fine-tuning the project. Selling a building where you deployed a solar array is an opportunity to monetize the present and future value the solar project creates for the underlying property. In these situations it is important to ensure that the solar array is recognized as an asset that has premium value, and not as an operational headache.

Solar can create value during the leasing process with tenants in the underlying building. The key is to identify the value it provides to tenants while clearly delineating who has responsibility for operational costs. In some situations this value may be direct, such as generating revenue by selling solar power to tenants. The benefit solar can play in attracting and retaining tenants may be more difficult to pinpoint, but it can be just as valuable if it leads to faster lease-up and higher tenant retention.

Part II of the handbook considers these key waypoints in the lifecycle of the solar array. You will learn to identify the ways that solar comes into play at these milestones. You will learn the steps you can take to ensure that the solar array is positioned as a building feature that creates direct financial value and supports an efficient transaction process.

Chapter 6

Solar for new and existing buildings

Chapter summary

- Incorporating solar into new construction projects allows you to exercise control over the design and aesthetics of the array, but adding the cost of solar to the project can create special challenges for project financing.
- Solar projects on new developments can help achieve points for sustainable building rating systems such as LEED, BREEAM, and others.
- Adding solar to existing buildings allows you to take into account the operating history, tenant profile, and solar incentive program guidelines to find the best host site.
- Solar on existing buildings may require sign-off from existing partners and lenders before a project can move forward.

Solar can be incorporated into buildings at many points in their lifecycle. Solar energy systems can be part of a new development project from its inception, or it can be added when tenants move in. Solar can also be added to a stabilized operating property. This chapter addresses the considerations that arise when incorporating solar at these points in the life of a commercial building. Taken together, these represent the most common situations you are likely to encounter when deploying solar across your real estate portfolio. Part IV of this handbook explores the considerations associated specifically with the design and construction best practices for incorporating solar into both new and existing buildings.

New construction

Solar can be incorporated into the planning and design phase of a new development or a redevelopment project. Including solar in the development of a new facility can have several advantages that are not available for solar projects on existing buildings. At the same time, there are potential pitfalls that can slow down a development project that pursues solar. Identifying both the opportunities to benefit and the possible risks better enables you to deploy solar on new development projects successfully.

The primary advantage of incorporating solar into a new development project is the ability to integrate the design and construction of the array into the overall project's

Table 6.1 Advantages and disadvantages of solar for new developments

Advantages	Disadvantages
Ability to integrate solar into architectural design	Higher design and construction costs than a project without solar
Ability to match the aesthetics of the solar array and building	Obtaining financing for solar portion of development
Savings due to shared project management and mobilization costs	Difficulty for the developer to capture solar tax benefits
Potential for improved community support due to positive perception of building with solar features	Unknown tenant demand for solar energy if tenants are not identified early in the design of the project
Potential to attract tenants seeking space in a sustainable building	Need to capture rebates and incentives from utility or government

design concept. This integrated approach streamlines the process in two ways. First, it allows the building design to be adjusted in order to make the solar installation easier and faster. Second, it allows the solar design to influence the building's design to ensure that the solar project can achieve its maximum production potential. This two-way interaction helps both the building and the solar array. Each can be designed and constructed more efficiently and cost-effectively.

There are several areas where this integration can benefit the development project. Integrating solar means that the building's designers can take into account the space needed for conduits and chases, electric rooms, and maintenance access to the location where the solar array will be installed. The structural design of the building can be adjusted if needed in order to accommodate the weight of the solar array.

There are three ways solar can be incorporated into the construction of a commercial building. Solar can be:

1   integrated into elements of the building façade or roof;
2   installed on a permanent structure added to the building;
3   added to the completed building shell as a separate, stand-alone system.

High-profile projects may support a solar array design that is integrated into building elements that are part of the façade or roof. This could, for example, be a solar array integrated into the curtain wall of the building, or it could be a courtyard canopy that has solar integrated into it as shading elements. In these cases the solar array is fully integrated into the design of the building, and removing the system would require deconstructing that portion of the façade. This type of fully integrated solar array is typically installed at the same time as the other components of the building are constructed. For example, the solar modules must be installed in the curtain wall in order to enclose the building and protect the interior from weather.

Solar can be designed to fit onto a supplemental structural frame that is added to the building specifically to support the solar array. An example of this is installing solar modules on a steel frame that is mounted above the entrance of a building to serve as an outdoor canopy. This allows solar to be added to the building with little impact on the rest of the construction process, while still having a high degree of control of the

Figure 6.1 A commercial building clad with an integrated solar façade (Source: photo courtesy of
Pete Birkinshaw, Manchester, UK, via Wikimedia Commons).

Note: This building, under construction in this photo, utilizes solar modules as the curtain wall façade of the
central tower. Removing the solar modules affects the appearance and integrity of the building façade in this
type of fully integrated application.

Figure 6.2 Solar modules on a steel frame of a façade (Source: photo courtesy of Friviere, via Wikimedia Commons).

Note: The curved solar canopy is attached to a steel frame that floats in front of the façade of the under-lying building. While the solar modules cannot be removed without affecting the property's appearance, their removal does not directly impact the function or integrity of the underlying building.

aesthetics of the overall design. In this case the building construction could proceed at its own pace and the solar modules could be installed at any point after the steel frame is completed. If there is a delay in installing the solar modules the rest of the construction work will not be impacted as significantly and the building itself could still be completed.

A third option is for the solar array to be a stand-alone system installed on the building, without the need for additional support structures or other visual changes to the building. This is the case where a solar array is installed atop the roof of a newly completed building. This provides less ability to customize the design of the solar array or to integrate it into the aesthetics of the building. Conversely it provides the greatest flexibility to install the system when the timing is right without impacting the completion of the underlying building construction.

This could be useful when the developer wants to sell solar energy to building tenants. A stand-alone system can be constructed when tenants take occupancy of the building without affecting the ability of the developer to finish the rest of the property. Because this type of solar project tends to be installed in less visible locations, it allows the array to be removed from the building at the end of its lifespan without negatively impacting the appearance of the building or requiring a costly replacement product.

Figure 6.3 An overhead solar canopy attached to the roof of a building (Source: photo courtesy of Marlith, via Wikimedia Commons).

Note: This building has custom-fabricated solar modules integrated into an overhead canopy. This canopy is permanent, but the solar array can be removed without affecting the integrity or function of the underlying building.

In communities that are sensitive to the environmental impact of real estate development activity, adding solar as part of a comprehensive sustainability strategy for the project can help with the permitting and approvals process. Solar can be a highly visible way to highlight broader efforts that have been undertaken at the property to reduce its ecological and environmental impact. In turn, these efforts may help to increase community support and show that the project developer is taking meaningful steps to minimize the project's perceived impact on the community and the environment.

Adding solar to a new development can be advantageous for marketing the building to potential tenants interested in a property that has environmentally sustainable features. In this case, making the solar installation highly visible will call attention to the features of the building. Consider using solar as part of the façade, a canopy near the main entrance, a covered carport, or integrated into the curtain wall of the building. In this scenario, the building should be designed to ensure that the orientation of the solar array faces toward the prevailing southern exposure where possible to help support higher production. If these high-visibility locations are not available due to the building's orientation or design, an alternative is to install a display kiosk in a high-traffic location, such as the lobby. This displays the output from the solar array and quantifies the environmental benefits of the clean energy it produces.

Figure 6.4 A solar array affixed to the roof of a building (Source: photo courtesy of the US Fish and Wildlife Service, via Wikimedia Commons).

Note: This solar array is attached to a standing seam metal roof. The array can be removed from the building without impacting the underlying building's appearance or function.

There are opportunities that come with incorporating solar into a new development, but adding solar is not always as simple as it seems. There are several reasons why solar is not more commonly incorporated into new development projects. These include:

- higher design and construction costs for the developer;
- difficulty obtaining financing for the solar portion of development;
- difficulty for the developer to capture solar tax benefits;
- need to capture rebates and incentives from utility or government;
- not having a tenant in place to purchase solar energy at the time of development.

Integrating solar into a new development project can enable more efficient construction and reduce project costs. A highly integrated, specialized solution can, however, actually increase construction costs far beyond an off-the-shelf solution. This may be justified if there is a value in a highly visible or customized solar array, although many development projects with limited budgets will find a more standardized solar approach preferable, even if it means the solar array is less visible.

Development projects often use outside sources of capital. While many capital sources leave the design decisions – including solar – to the developer, there are cases where solar adds a meaningful amount to the project cost. This could raise flags with the third-party capital provider if the value proposition of solar is not clear to them. Fortunately, purchasing a solar array outright is only one of several popular solutions for procuring solar. Cost-effective solar financing solutions exist outside of simply rolling the cost into the development budget. These solutions, referred to as *project structures* in this handbook, are discussed in detail in Part III.

The capital to fund the solar portion of the development may be better provided by a different entity than those supplying the development financing. Adding a solar financing package on top of the development financing will add complexity to the project, but keeping the two projects financially distinct can make each one easier to execute. Keeping each financing solution within the standard expectations of each provider can be simpler and more efficient.

Where third-party solar financing is needed, recognize that their financing is typically only available to solar projects that are stand-alone projects connected to the building where there is a long-term energy purchaser. Projects heavily integrated into the building are difficult to finance since the line between the solar array and the building cannot easily be established. This makes it difficult to delineate what the collateral is for the financing. A self-contained solar array that sits atop a roof could readily be financed. An array integrated into the curtain wall of the building façade could expect to have difficulty finding solar financing.

In solar markets where tax-based incentives are available, the developer may find that the entity they have created to develop and own the property is not equipped to efficiently capture the solar tax benefits. These entities are often limited partnerships or limited liability corporations. These legal entities pass through tax liabilities and benefits to the partners or members, so there is usually little ability for the entity itself to take advantage of solar tax benefits. Putting a structure in place that enables tax incentives to be captured can make the development structure more complex than would otherwise be required if solar was not being considered.

In markets where utility-funded solar incentives exist, a new development project may find it has difficulty applying for rebate programs simply because the building – and its associated utility account – does not yet exist. In some cases a customer's utility account will have to be in good standing for a pre-determined period of time in order for the property owner to become eligible for incentives. On a new development this may not be possible until many months after the project has been completed. In this case it may be easier to plan for the solar project but hold off on the installation until the property is eligible for utility rebates.

Developing a project with solar can have an added dimension of financial risk if the developer expects to sell solar energy to tenants. Many speculative development projects commence without tenants in place. This means the developer could face a situation where they are unable to sell solar energy to tenants to recoup their solar costs even though they have already invested the capital to purchase and install the solar array. This risk can be minimized by using the solar energy to power common area energy needs that will exist regardless of whether there are occupants in the building. This risk can also be managed by installing the solar array after tenants have taken occupancy of the building.

> **Lessons from the field**
>
> *Sustainable building certifications*
>
> More and more development projects are pursuing certification under sustainable building ratings systems. These rating standards, such as the LEED rating system from the US Green Building Council, BREEAM, CASBEE, and others offer ways for developers to receive credit for incorporating solar into their projects.
>
> The LEED rating system offers credits for producing a portion of the building's energy on-site from renewable sources such as solar. The rating system developers acknowledge that solar energy enhances the sustainability of the underlying building and reflect that in their scoring systems.

Existing buildings

Many existing buildings are potential host sites for solar projects. The vast majority of solar projects are built on existing properties, not on projects under development. This is for several reasons. There are many times more standing buildings than there are construction projects in any given year. In most markets this new supply equals no more than 1–2 percent of the existing building stock.

Many property owners intuitively look to their operating portfolio instead of new developments when considering solar for the first time. This is because there are fewer moving parts and less uncertainty to developing a solar project on a stabilized asset. There is also less time pressure. Since the building is already standing, there is less need to coordinate the solar work with other construction activity. Especially on their first few solar projects, property owners do not want to be rushed before they can consider all the factors that contribute to a successful installation.

Existing buildings offer the ability to match solar with the operating profile of the property. Unlike a new development, an existing building will have years of operating data – energy use, roof condition, knowledge of tenants' interest in solar, and the like. This information makes it easier to survey the options for deploying solar. It can also

Table 6.2 Advantages and disadvantages of solar for existing buildings

Advantages	Disadvantages
Extensive operating history available for building	Limited ability to integrate solar into building design
Easier to identify potential sites that fit solar program requirements	Older buildings may have physical limitations for supporting a solar array
Wide range of financing solutions tailored to solar on existing buildings	Potential disruption to building operations and tenants
Ability to coordinate timing of solar installation with other capital projects	May need to notify mortgage holders, joint venture partners, co-owners

help direct the property owner to the best solar solution for the building. On a new development project all this information is either assumed or based on engineering projections, and therefore subject to greater uncertainty. For example, depending on the type of tenants that eventually occupy the building, energy needs could be quite a bit different than those modeled during the design.

Existing buildings offer more options to match up with solar incentive programs in a given market. In many solar-friendly markets, financial incentives for solar projects are tailored to projects of a certain size. The owner of a portfolio of existing properties can select the property that is the best fit for the incentive program, while the developer's new construction project may not be a good fit for the same program.

For example, consider a solar incentive program that is tailored for projects of approximately 100 kW; a property developer's new project may not have sufficient available space to support a solar array of this size. Conversely, a property owner managing a portfolio of properties can sort through their assets and select the property that will readily support a 100 kW installation.

Adding solar to existing buildings enables the property owner to coordinate capital projects to coincide with the solar project. For example, a roof replacement is often completed immediately before solar is added to a building. Repaving a parking lot and re-landscaping could be coordinated with the installation of a solar parking lot canopy. While this type of coordination is not required, it better aligns the useful life of building components with the lifespan of the solar array. It can also reduce the re-work that would otherwise be required if the projects were not coordinated in this manner.

At the other end of the spectrum, adding solar to an existing building means you will be dealing with the building in as-is condition. This may mean that the solar array cannot be as large as you would like it to be, or that the solar layout is less efficient because it had to be designed around existing building features or equipment. On older buildings, structural limitations may necessitate design changes that increase costs or reduce the area available for solar. Zoning or other regulatory requirements may also place restrictions on the solar array.

While a building full of tenants is good for the property, the reaction of tenants to solar can vary. Many see it as an opportunity to become more sustainable, but some could see the possibility of unwanted disruptions. This could include concerns about noise, parking limitations, the risk of leaks, or potential interruptions to electric service. A well-planned project can manage these potential impacts effectively, but concerns expressed by building tenants are important to weigh when deciding which sites are the best candidates for solar.

Adding solar to an existing building means that lenders, limited partners, insurers, or other parties with a financial interest in the underlying building should be notified of the planned project. Depending on your relationships and the terms of contractual agreements, this may be a simple advisory call, while in other cases the counterparty may have the right to decide if the project can move forward. For example, a mortgage lender may have the right to review and veto major alterations to the property, such as adding a solar array. Executed properly and communicated to counterparties clearly, solar adds value to properties and is therefore often looked at in a positive light. It is important to recognize that the added step of working with existing parties in the property's ownership and financing structure can add time and additional decision-makers.

Final thoughts

Solar can be incorporated into a commercial building at various points throughout its lifespan. Planning solar as part of the development process offers advantages such as efficiency of construction and an aesthetically integrated solar array. Solar can also be incorporated onto a property months or years after development is complete. Adding solar to existing stabilized assets is one of the most common ways solar is incorporated into commercial buildings. In either case, understanding the benefits and trade-offs will help ensure that the project creates value for the property developer and building owner.

Chapter 7

Solar in acquisitions, dispositions, and leasing

Chapter summary

- Acquiring a property that already has a solar array is an opportunity to capture the benefits of solar without the need to develop it.
- Selling a building that has solar is an opportunity to capture the value solar creates for the asset.
- Leasing space in a building with solar means identifying the value it provides to tenants while clearly delineating who has responsibility for operational costs.

Solar can play an important role when buying and selling commercial buildings and when leasing space in a building. When marketed effectively, the solar array on a building has the opportunity to create value for both the seller and the buyer of the property. To ensure this value is realized the seller should provide accurate information to the potential buyer so they can quantify its value in their offer price. Because few buyers have experience assessing the value of solar during the acquisition process, providing complete information enables the buyer to fully understand the solar array and recognize the value it creates. Without this understanding, the seller risks that the buyer either ignores the value of the solar array or mistakenly counts it as a liability.

In a similar manner, leasing a building with solar to tenants should be approached in a way that demonstrates the value solar can provide to lower energy costs and reduce the carbon footprint of the tenant's operations. This helps to ensure that the value of solar is reliably captured for both the landlord and the tenant.

Acquisitions and dispositions

Acquiring a property that already has a solar array is an opportunity to capture many of the benefits of deploying solar without the need to develop the solar array. Care does need to be taken, however, to ensure that the array meets expectations and does not become a liability in the future. Focus on the following due diligence areas:

- Read and verify all solar project contracts to ensure you understand what obligations and responsibilities you will be taking on. Common contracts include power purchase agreements (PPAs), roof leases, tenant solar lease amendments, power re-sale agreements, and renewable energy certificate (REC) sale contracts.

- Verify whether there are extension options on contracts such as roof leases or power purchase agreements.
- Identify financial obligations associated with contracts, such as early termination fees, buy-out clauses, and repair cost obligations.
- Request a performance history of the solar array to see how it has performed, both in terms of financial performance and electrical output.
- Review financial statements and payment histories for solar accounts. This could include payments to a solar company, as well as invoices and receipts from tenants, the utility, or another energy purchaser. Verify whether there are any utility payments or tax incentives outstanding and review the process for collecting them.
- Request warranties on solar equipment and verify that they are still valid.
- Have an experienced solar operations and maintenance contractor inspect the system and the performance data to determine if it has been maintained properly and is performing as expected. Have them estimate the remaining useful life of the system and key components such as inverters.
- Cross-reference the remaining expected life of the solar array with the expected life of the adjacent building elements. If the system is on a roof, check how much remaining life the roof has. If these are not aligned, consider the potential costs.
- Look for signs of potential environmental impacts that may need to be addressed, such as a leaking transformer or broken solar modules.

Reviewing these topics will provide you with the key information needed to model the financial performance of the solar array and assess its value in the acquisition process. You may want to have any major problem areas corrected as a condition of the sale, but it may be easier to discount the value of the array to account for this scope of work. If you have time in the transaction, you may want to determine whether there is potential to expand the solar array to accommodate future energy needs or to increase revenue.

When selling a property that has a solar array, expect that the potential buyer will ask for the same information described in the section on acquisitions. Because few buyers have extensive experience with solar projects, having this information readily available will help them understand the project and its value. The more difficult it is for the potential buyer to assess the quality and value of the solar array, the longer it will take them to get comfortable with the idea of purchasing a building with solar.

Leasing

Leasing a building that has a solar array can be looked at from two perspectives. The first is that of the landlord seeking a tenant to occupy the building and purchase the electricity produced by the solar array. The second perspective is that of the prospective tenant seeking space in a building. In any transaction the goal should be to find a common ground where both the landlord and the tenant can benefit from the lease. Solar offers a way for the landlord to benefit by:

- generating revenue from the sale of energy from the solar array;
- attracting tenants to the property;
- providing an amenity that supports renewal of existing tenant leases.

Tenants can benefit from solar by:

- reducing energy costs;
- supporting corporate sustainability goals – capturing solar RECs (sRECs) from the use of solar to reduce their carbon footprint;
- demonstrating environmental responsibility to employees and the community.

Solar can attract tenants to the building and differentiate the property from others in the market. Solar can also be a visible symbol to customers of the commitment to operating responsibly and reducing the impact on the environment. For many existing tenants solar is an attractive amenity that encourages them to lease or renew space. Of course, not every tenant is going to have an interest in solar for its environmental benefits. For them, the solar project may be of value as a way to save on energy costs by purchasing solar energy at a discount to utility-provided electricity.

Landlord perspective

A tenant interested in solar may seek a building that already offers solar energy. Alternatively, they may seek a property that has the potential for them to construct their own solar array. Depending on the perspective of the tenant, the property owner can position their property accordingly. This may include:

- highlighting in marketing materials that solar power is available on-site, including the amount or percentage of energy it can supply, e.g., "Suite 100 can be powered by 100 percent solar energy";
- providing a pro-forma calculation that demonstrates the cost savings from purchasing on-site-generated solar energy;
- noting the area of the property that is "solar ready," such as the building's roof or an open area reserved for solar on the ground – this could include the potential size of the system and an example solar layout;
- highlighting the way solar supports sustainable certification for the tenant's space, e.g., LEED certification for interior tenant spaces.

These aspects of the project will be one of many considerations in the potential tenant's leasing decision. But for the right customer, highlighting the value of solar is a way to differentiate your property and build a convincing case that solar provides long-term benefits that cannot be matched by another building.

Tenant perspective

The perspective of most tenants revolves broadly around two themes: (1) What is the benefit solar provides them? (2) What are the risks or obligations they will be exposed to? It is important to address both areas proactively. The benefits to the tenant commonly include:

- lower cost of electricity;
- more stable electricity prices;
- ability to claim their property is powered by clean solar energy.

Of these potential benefits, expected cost savings is often the first area to focus on because it is the most quantifiable. As a related benefit, tenants may recognize value in stabilizing energy costs throughout the term of their lease. One way this can be accomplished is by having a contract for tenant solar purchases that is indexed to the retail price of electricity, less some percentage.

Another area that tenants may be interested in is the ability to state that their facility is powered by clean, emission-free solar power. This can be a powerful message to the company's employees, many of whom may be gratified to see their employer supporting environmentally responsible sources of energy. Taken as a whole, the financial and reputational benefits of solar provide a strong incentive for tenants to seek out buildings that offer solar.

Expect the tenant to have questions about the potential benefits of the solar array. Common questions relate to:

* risk of leaks or other potential damage to tenant property;
* possible increased building operating costs;
* ensuring that they are getting the cost savings they were promised;
* whether solar contractual requirements will add complexity to the lease.

Tenants may be concerned that the solar on the building, particularly when on the roof, may lead to an increased potential for leaks that could damage the tenant's space. This can be addressed effectively in most cases where the roof is new and its lifespan is in line with that of the solar array. If that is not the case, the landlord needs to be clear about which party is responsible for building maintenance and leak repair in the areas where the solar array is located.

Tenants may be wary of the potential for solar to increase building operating costs. This could be due to a higher number of inspections, repairs, and other upkeep related to the solar array – costs that would not be necessary at a building that did not have solar. To address this, the landlord should be able to clearly define the expenses the tenant will be responsible for, including any foreseeable solar maintenance, if applicable. In general, tenants are unwilling to pay for solar maintenance costs unless they are receiving a significant offsetting financial benefit from the solar array.

Tenants also focus on ways they can guarantee the electricity cost savings they have been promised. This can be done by establishing a consistently measurable savings benefit for the tenant, such as indexing their payments to their retail electricity price minus some discount. Minimum production requirements for the solar array operator can also help get a tenant comfortable with the expected quantity of solar electricity. Sharing data on historic and expected production provides confidence regarding what they can expect. By de-risking the tenant's participation in a solar transaction in this way, the value created by solar will become clear.

Another concern that can arise for tenants – and landlords too – is that solar can add complexity to the lease transaction. There are two approaches to this. One is for the solar agreement to be incorporated directly into the lease negotiation, usually through an amendment to the lease. The second is for the solar agreement to be a stand-alone document that is negotiated separately from the lease itself. Keep in mind that the solar component of the lease transaction is only a small portion of the value of the tenant's lease. This perspective helps to avoid making the solar discussions a bigger issue than they really are. Keeping this in mind allows solar to be an enhancement to the lease rather than the focus of the negotiation.

Final thoughts

Throughout a building's operational lifecycle – acquisitions, dispositions, and leasing – solar can make the building more attractive to buyers and tenants. The result is increased value of the underlying property. At these milestones in the property's life it is important to clearly identify the ways in which a solar project benefits both the property owner and the other party in the transaction. This provides a clear path to finding common ground and recognizing the value solar provides.

Part III

Selecting the best project structure

The three chapters that comprise this section explain the three primary project ownership structures that may be available to choose from in the market in which you are pursuing solar. From the perspective of the property owner, a solar project structure is the legal and financial ownership arrangement by which the project is developed, owned, and operated. There are three predominant solar project structures described in this handbook:

- *Direct ownership* is a structure where you purchase and own the solar project yourself.
- A *power purchase agreement* (PPA) is a structure where a third party develops and owns the solar project and sells electricity to you under a long-term contract.
- A *lease* is a structure where a third party leases space on your property to develop and own a solar project. They sell electricity to the utility or an unrelated third party under a long-term contract.

I refer to the three structures respectively as "ownership," "PPA," and "lease." These structures allow you to package all the necessary elements – location, incentives, technology, and electricity prices – into an efficient financial structure to create a profitable project that is tailored to the particular needs of your business and your project. Knowing what project structure works best for you allows you to seek out solar service providers that have the expertise to plan, finance, and deliver that type of project. Table PIII.1 provides an overview of the three structures and the major differences between them from the perspective of a range of project objectives. Table PIII.2 defines the responsibility for various aspects of the project for each project structure. Use these tables to begin matching your project's goals with the structure that is most appropriate for your needs.

As you review these tables you will begin to identify the project structure that is likely to meet your solar project goals. You can identify specific criteria that may drive you to one structure over another. For example, your goal with solar may be to offset common area electricity consumption, but you may not have capital available to invest in a solar array. Based on the preceding tables you would likely find a PPA the most suitable for your needs. Another property owner may decide that they only want to use their roof space for revenue-generating purposes and they care little about offsetting building energy use. In this case a lease is likely to be the best solution.

Each of the three project structures exposes you to certain risks that you need to recognize and be comfortable with. Adding a solar facility at your property can impact

Table PIII.1 Objectives met by each project structure

Project structure	Objectives
Direct ownership	• Offset common area energy needs or sell power to tenants • Control design and operation of solar project • Capture any financial upside from solar investment • Own solar environmental attributes, including RECs and carbon emission reductions
Power purchase agreement	• Offset common area energy needs or sell power to tenants • Avoid out-of-pocket solar expenditures • Limit involvement with solar design and construction
Lease	• Generate incremental revenue • Avoid out-of-pocket solar expenditures • Limit involvement with solar design and construction

Table PIII.2 Responsibility within project structures

	Development costs	Solar O&M	Control solar RECs	Use solar electricity
Direct ownership	Property owner	Property owner	Property owner	Property owner
Power purchase agreement	Solar developer	Solar developer	Solar developer	Property owner
Lease	Solar developer	Solar developer	Utility company or third party	Utility company or third party

Table PIII.3 Relative risks to the property owner

	Financial risk	Development risk	Operational risk
Direct ownership	High	High	High
Power purchase agreement	Medium	Low	Low
Lease	Low	Low	Low

the building, its operations, and the revenue available to you. For example, purchasing and constructing a solar array under the direct ownership structure presents a very different, and greater, risk profile than leasing a portion of your rooftop to a regional utility that owns and operates a solar array that supplies their electricity grid. Table PIII.3 illustrates the relative range of risk for each project structure.

There is no "better" structure compared to the others. In fact, given the right set of circumstances, each structure may work for a property owner at a different time or property. Within each structure, there are a number of variations that can be adapted to suit particular needs. For example, rather than fund the construction of a solar array, you may want to have a third-party develop the solar array, and even operate it for several years, before having the ownership revert to you at an agreed-upon price. As you read the characteristics for each structure in the paragraphs and chapters that follow, think about which structure is likely to be best-suited for the assets in your property portfolio.

Any of the project structures can, if not handled correctly, affect existing contractual relationships you have with building tenants, lenders, investment partners, and other stakeholders. You may find that it is faster and easier to pursue one project model over another because of the approval requirements of stakeholders that already have contractual relationships with you at the property. While signing a PPA may require only minimal stakeholder input, directly investing in a solar project may require in-depth review and approval from partners, lenders, and others with an interest in the underlying property. As a result, they should be advised of the solar project and allowed sufficient time to review or approve the project as needed. This will help ensure they do not become a potential stumbling block once the project is underway. Even if stakeholder approval is not required, an early and open dialogue will help gain their support for your solar project, regardless of the project structure you choose.

How to use this section

Even at this early point before reading the next few chapters, you may begin to have a sense for which project structure will best fit your objectives. If so, feel free to skip ahead to that chapter. The handbook covers direct ownership in Chapter 8, PPAs in Chapter 9, and leases in Chapter 10. I do, however, recommend reading each of these chapters in order to become familiar with the considerations that affect each project structure.

Because solar incentive programs and regulations vary from one region to another, this knowledge will enable you to select the structure that is most appropriate for a given property in any market. Keep in mind that it is possible and even likely that over time more than one solar project structure could work for you. By the time you are finished with this section of the handbook you will have a solid understanding of the opportunities and trade-offs that come with each of the three project structures. This will enable you to select the one that allows you to manage risks effectively while meeting your financial and operational goals.

Chapter 8

Direct ownership

Chapter summary

- Under the direct ownership structure, solar electricity is typically used by the building to power common areas, or sold to building tenants.
- Direct ownership provides the greatest income potential but also the greatest risk of lost revenue and capital costs.
- Direct ownership allows property owners to control the renewable energy certificates (RECs) generated by their project.
- Solar developers or engineering, procurement, and construction (EPC) companies may be used to provide turnkey project delivery services.

Of the three basic structures for solar agreements, the most straightforward is direct ownership, where you purchase a solar array from a company that handles the design, permitting, and installation. Once the project is complete and operational, you own it and can do with it as you see fit. You are responsible for maximizing revenue and for capturing incentives and tax benefits. You receive the income generated from the sale of electricity or RECs produced by the array. You are responsible for managing the operational impacts the array has on your property, both today and into the future. Overall, direct ownership offers the greatest level of control over the solar project for property owners that have the interest, time, and capital to pursue this structure. The direct ownership structure is usually the only structure suited to projects that have a highly specialized design that is integrated into the building, such as solar glazing integrated into a curtain wall. In these cases it is unlikely that a third party will be willing to finance and own this type of installation via a power purchase agreement or lease.

Revenue

Owning a solar array outright typically has the greatest recurring revenue opportunity of the three structures. Your project may receive some or all of the following financial benefits, subject to availability for your particular project:

- cash incentives;
- tax credits, deductions, and depreciation;

- revenue from selling solar electricity or from offsetting your utility bill;
- revenue from selling RECs if available in your market.

To capture these benefits, you need to have a financially efficient project ownership structure. To understand this aspect of direct ownership projects, ask yourself the following questions:

- Do you have a cost-competitive source of internal capital to fund the design and construction? This may include debt as well as equity. If you do not have a cost-effective source of capital, you may be better off pursuing a different project structure such as a PPA.
- Do you have the resources and knowledge to apply for available solar incentives? Depending on the market, these applications can be fairly complex and may require both an initial application and a final submittal that has sign-offs from multiple parties, including local inspectors, utility representatives, and the installer. In some markets, the disbursement period for incentives can span several years and requires annual reporting in addition to the application itself.
- Is the legal entity that owns the solar project set up to efficiently capture tax benefits that accrue to the solar array? This can be a challenge for real estate companies that are structured to have low tax liabilities. An example is the real estate investment trust (REIT) that is exempted from most corporate taxes in the United States. Some REITs use taxable subsidiaries as a repository for the tax benefits. Limited liability corporations (LLCs) and limited partnerships (LPs) may also have difficulty monetizing tax benefits.

If the answer to one or more of these questions is "no," it may be difficult for you to pursue the direct ownership of your solar project. This is particularly true if you do not have a tax-efficient structure because of the importance of being able to monetize these valuable incentives. Table 8.1 summarizes key considerations for capturing tax benefits from solar projects for several common real estate ownership structures.

Electricity value

Electricity produced by a solar array can be monetized in one of three ways under the direct ownership model:

Table 8.1 Factors that affect the suitability of direct ownership

REITs	LLCs and LPs
• Tax credit needs to be claimed by taxable subsidiary • Taxable subsidiary may not have tax appetite in current year; carrying tax benefits into future years diminishes their value • Solar revenue may count as non-qualified (i.e. "bad") REIT income	• Tax credit need is passed through to members or partners • Members or partners may not have tax appetite in current year; carrying tax benefits into future years diminishes their value • Each member or partner may place different values on tax benefits

1   offsetting owner-controlled building energy use;
2   selling the electricity to building tenants;
3   selling electricity to the utility under a feed-in tariff contract.

Electricity used to offset owner-controlled energy demand is most often used to reduce common area energy consumption or other "house" meters where the property owner has responsibility. Electricity sold to a tenant is typically priced at some negotiated percentage less than its total retail cost. This might be a rate equal to 90 percent of their current electricity price, so if the utility rate is ten cents per kilowatt-hour, the cost of solar electricity would be nine cents. This below-market price provides an incentive for the customer to purchase solar electricity from the project to supplement existing grid-supplied electricity.

The price of solar electricity could be set to rise as utility electricity prices increase over time, so it remains the same percentage below the then-current utility price. Alternatively, you might sell tenants solar electricity at a price equal to their current utility price, but fix the price so it stays flat even as utility rates rise in the future. As there is tenant turnover you could reset the rate you charge back to the current price of electricity for a new tenant. There are many other variations, such as stepped prices that reset to market rate every few years, or a fixed price that converts to a percentage rate based on the market price after several years.

In some markets solar electricity can be sold to the electric utility. This arrangement is common in many parts of the world through FiT contracts. With a FiT you receive a payment for all electricity delivered to the grid. In some locations you are required to offset your own energy use before putting energy back on the grid. This can create both lower operating costs as well as a revenue stream from any excess energy produced. In the United States, FiTs are much less common than in other parts of the world. In most cases in the United States, the utility will not pay you if you produce more electricity than you consume. Instead, you will get a credit against future energy use on your bill. This credit is often applied to your total bill, so it does not take into account when the energy was produced. While you may generate excess electricity when grid prices are highest, the utility may apply the credits to your account when prices are much lower. There are several exceptions to this situation, such as in southern California where solar can be sold to the utility by participating in a competitive bid program that they administer.

Financial model

The direct ownership structure requires you to develop the financial models that underpin the project. You can utilize the expertise of your solar service providers to help construct this model, but you will want to verify any assumptions in their models with your own experience and financial situation. For this reason it usually makes sense to have your own financial analysis instead of relying on a solar developer or other service provider to supply one. This financial model will help you determine the best project solution to maximize revenue. This financial model can help you answer a number of questions, such as:

• Is committing funds to solar an attractive use of capital?
• Is a more expensive, high-efficiency solar module better than a less efficient commodity module?
• What is the impact of receiving tax benefits this year or in future years?

- How does debt affect returns?
- Does the discount rate your solar provider used in its model match your own?

These are a few of the many questions you will be able to answer when you have developed a solar financial model. In addition to looking at the solar project as a stand-alone investment, you may want to include it with other capital costs such as an energy efficiency upgrade, retro-commissioning, or a re-roofing project. This holistic approach may allow you to cross-subsidize the package of improvements to allow you to accomplish more within a single project.

While you do not need to rely exclusively on solar service providers for financial model inputs, they are a good resource for baseline data and assumptions for you to initially populate your model. You can begin with inputs from proposals provided by solar companies or service providers and adjust assumptions as you develop your own financial model. Talk to several service providers to help identify a realistic range of costs and revenue assumptions. Obtaining several proposals for solar projects will give you insight into what project costs you can expect. This information will be useful for the direct ownership structure, as well as the PPA structure and the lease structure.

To construct a financial model, you can either begin with one you use for real estate projects, one provided by a solar company, or one you create from scratch. Set up the model based on the expected operational life of the solar facility – typically 20 years. Add cost line items as shown in the example financial model illustrated in Table 8.2. Remember to factor in inflation for recurring expenses such as operation and maintenance (O&M) costs.

These costs will be offset by the sources of revenue shown in Table 8.3. The exact mix of revenue sources will depend on your market, available incentives, and what you plan to do with the electricity and RECs. Some of these may also not apply to your project, depending on where it is located and whether there is value in the RECs you create. In some markets the value of RECs is sufficiently low that it is not cost-effective to sell them.

Add to the financial model the rows shown in Table 8.4 to capture tax-related benefits that may be available for the solar project. Some of these, such as tax exemptions, are

Table 8.2 Financial model costs

Costs	Year 0	Year 1	Year 2	Year 3	...	Year 20
Construction costs including contingency	−950,000	−	−	−		−
Grid interconnection costs	−5,000	−	−	−		−
Engineering costs	−20,000	−	−	−		−
Permitting fees	−2,000	−	−	−		−
O&M costs	−	−1,000	−1,030	−1,060		−1,754
Replacement reserve	−	−800	−800	−800		−
Legal fees	−8,000	−	−	−		−
Broker fees, if any	−	−	−	−		−
Financing costs	−	−	−	−		−
Vacancy loss on electricity sales to tenants	−	−1,000	−1,030	−1,060		−1,754
Subtotal costs	−985,000	−2,800	−2,860	−2,920		−3,508

Table 8.3 Financial model revenues

Revenue	Year 0	Year 1	Year 2	Year 3	...	Year 20
Cash grants	6,000	–	–	–		–
Utility rebates	10,000	–	–	–		–
REC sale value	–	10,000	9,500	9,025		6,500
Electricity savings	–	–	–	–		–
Electricity sales revenue	–	75,000	77,250	79,568		131,513
Subtotal revenue	16,000	95,000	102,250	88,593		137,013

Table 8.4 Financial model tax benefits

Tax benefits (costs)	Year 0	Year 1	Year 2	Year 3	...	Year 20
Federal tax credit	–	285,000	–	–		–
State tax credit	–	–	–	–		–
Sales tax exemption	12,800	–	–	–		–
Property tax exemption	–	9,500	9,500	9,500		9,500
Deductions	–	–	–	–		–
Depreciation	–	57,000	91,200	54,720		–
Taxes on solar rebates	–7,800	–	–	–		–
Taxes on electricity revenue	–	–	–	–		–
Taxes on REC revenue	–	–	–	–		–
Subtotal tax benefits	5,000	351,500	100,700	64,220		9,500

avoided costs, not revenue. They are included in the financial model for purposes of giving a complete picture of the sources of value accruing to the project.

It is important to model the timing of major cash flows accurately, particularly in markets that incentivize solar with cash grants. When your project is counting on lump-sum incentive payments to offset construction costs, you need to have a good idea of when you can expect the income. In some cases, incentives may be paid upon project completion. Find out how long it will take to get paid – it could take months for your request to be processed. In others they could be paid annually in arrears and spread out over several years. This also applies to tax credits and other tax benefits that you may not be able to monetize immediately. Being able to accurately model major cash flows helps create a realistic representation of the financial performance of your solar project. Consult with your tax advisors to quantify when and how tax benefits can be monetized. Ask solar companies and other service providers how long it will realistically take to get available cash grants or other incentives for your project. The processing time for solar incentives can vary significantly from one market to the next.

For projects that are intended to sell electricity to tenants, your financial model may need to account for the possibility that a tenant space could become vacant during the operational life of the solar project. The departure of an electricity purchaser could limit your ability to sell electricity. This has the potential to reduce solar revenue during the period of the vacancy. One way this scenario can be minimized is to design the solar array to be a size that is unlikely to produce excess electricity even during periods of partial vacancy and diminished demand. By designing the solar array to meet a portion of the solar electricity needs of several tenants in the building, the loss of any one tenant will still leave you with enough energy demand to avoid losing revenue.

Quick tip

For quick feasibility analysis it is often not necessary for your financial model to capture every detail of the project perfectly. At the point when you are comfortable that a project is viable and worth pursuing you can develop an "investment grade" financial model that has the detail you need to make an informed decision.

Renewable energy certificates

Under the direct ownership structure, the project sponsor retains control of the RECs that represent the clean energy attributes of the solar electricity. This provides flexibility to monetize them by selling them if they have value. They can also be retired; counted toward corporate clean energy purchasing goals or carbon reduction commitments. Be aware that if you sell electricity to a utility under a FiT contract, you will in most cases be obligated to bundle the RECs with the electricity in exchange for the financial support provided by the utility.

Because the clean energy attributes of the solar project are controlled by the project sponsor, there may be marketing or other public relations benefits that could create value for the project. These could come in the form of enhanced reputation as a responsible company in the eyes of customers, investors, or the local community. These benefits may increase community support for other projects and make it easier to get regulatory approvals in the future.

Risk profile of direct ownership

Direct ownership provides an opportunity to capture a significant new source of revenue, where any income associated with the solar array flows back to the property owner. In exchange for this potential revenue, the direct ownership structure carries risks for the property owner at major stages of the project's lifecycle. This part of the book provides a brief overview of the most common risks associated specifically with the direct ownership project structure; in Part IV you will find detailed coverage of the range of risks that apply broadly to nearly all solar projects. For the direct ownership structure, risks commonly include:

- cost over-runs during design, procurement, and construction;
- difficulty obtaining financing or capturing incentives;
- inability to use or sell electricity at the price and quantities anticipated;
- higher-than-expected costs of ongoing operations and maintenance;
- decommissioning and disposing of the project at the end of its lifespan.

Under the direct ownership structure, the property owner is the project sponsor, and is responsible for costs associated with the construction of the project, including cost over-runs, delays, or other problems that may arise. Some of these risks can be mitigated with careful due diligence and proper design of the solar array. Construction contracts can shift certain project delivery risks to contractors and vendors. Not all risks can be mitigated entirely, however, leaving the property owner with a share of the potential risk to shoulder if the project does not proceed smoothly.

The project sponsor is responsible for arranging funding for the project. This funding may include cash, debt such as a construction loan or equipment lease financing, solar incentives, and likely a combination of several of these. Changes in the availability of capital or its cost are risks that may occur, and may not be entirely controllable. For example, failure to obtain debt at a favorable rate may result in higher project costs or it may lead the project sponsor to have to invest more cash into the project than originally expected.

A key requirement for direct ownership projects is the ability to utilize or sell the electricity produced by the solar array. Most commonly, the energy is used to offset building energy needs or sold to building tenants under a long-term power sales contract. Regardless of what you do with the electricity, there is risk that over the lifetime of the project there could be times when the electricity cannot be sold or used profitably. A tenant leaving unexpectedly could leave you temporarily without a customer for the electricity. This risk can be managed by incorporating periods of vacancy or other disruption into your financial models. Signing power sales agreements with strong-credit tenants or multiple tenants can reduce the impact of losing a solar electricity customer. Sizing the solar project so it produces no more electricity than you can use or sell even under periods of reduced demand can also reduce this risk.

Costs of operation are the responsibility of the project owner under the direct purchase structure. Increases in O&M costs can erode project returns, particularly if there are unanticipated costs, or if costs rise faster than expected. Failure of costly electrical equipment such as an inverter or transformer are the most likely causes of significant increases in costs, but higher-than-expected costs to clean and maintain the system can also add up over time.

Similar risks exist at the end of the life of the solar array. Twenty or more years into the future, it is difficult to predict what the cost of removing and disposing of solar equipment will be. The costs could be quite low and there could even be a secondary market for used solar modules in the future. Conversely, the cost of disposal may have increased greatly from where they stand today. This risk cannot be entirely mitigated, but budgeting conservatively for these costs and re-evaluating the likely cost as you approach the end of the project's operational life will allow you to manage it effectively.

The direct ownership structure occupies the highest position on the risk and reward spectrum of the three project structures. This risk and reward proposition may be attractive for experienced property developers and property owners accustomed to taking calculated risks. For an asset manager focused on maintaining returns at a stabilized property, this project structure may have too much risk because it lies outside their core investment strategy. Depending on your perspective, direct ownership may be an exciting opportunity or it may be seen as a risky option compared to the other solar project structures.

Cost considerations

As noted previously, the costs of developing a solar project under the direct ownership model will, directly or indirectly, be borne by the project sponsor. These costs include:

- site visits and conceptual design of the system;
- engineering the solar array;
- solar module and other system equipment specification;
- permitting;
- roof preparation;
- material procurement;
- site preparation;
- installation;
- interconnection to the utility grid;
- system commissioning;
- paperwork required to capture solar incentives;
- legal and tax advice;
- overhead costs.

For most solar projects on commercial buildings, working with an experienced solar development company is an expedient solution for project delivery that bundles many project costs. Their specialized experience managing solar design, permitting, and construction allows them to fill in many of the most common gaps that real estate property owners have when it comes to deploying solar projects.

EPC contracts

An EPC contract includes both hard and soft costs. Hard costs for solar modules and other equipment typically compose more than half of the total project cost. Solar modules are the largest component of hard costs. The remaining soft costs include construction labor, engineering, permitting, and overhead. Also include a line item for profit. Exact percentages for any given project will vary based on a number of factors, including:

- location;
- project size;
- labor costs;
- technology;
- utility interconnection costs.

These costs can be procured as a turnkey package from a solar developer or, depending on the level of control desired, you could purchase them à la carte. Particularly for your first few projects, a solar developer's expertise will be a valuable asset in your efforts to successfully manage project risks and deliver a successful project. As you gain more comfort and experience, you may choose to limit the services provided by a solar developer if you feel comfortable taking on additional aspects of project development.

Timing of solar incentives

Under the direct ownership structure, solar projects are often funded from the balance sheet of the project sponsor. Due to the short construction time frame, construction loans are not widely available in most markets for commercial-scale projects on buildings. Cash rebates and other incentives are generally only available after projects have become operational. You may not see all of your incentive dollars for up to a year after construction ends, so this source of funds cannot be counted on to pay suppliers or contractors.

There are solar equipment lease financing solutions that can help property owners minimize their out-of-pocket solar costs. A growing number of banks offer this financing product. Some solar installers may also offer solar leasing through their financial partners. These leases generally have terms ranging from seven to fifteen years. Because solar equipment cannot readily be repossessed by the bank at the end of a lease, contractual provisions generally require a buy-out at a specified price at the end of the lease term. With more lenders offering lease financing it is poised to become a more common option for commercial property owners.

Capital expenditures

Under the direct ownership structure you may want to include a reserve to cover capital costs and the cost to remove the solar facility at the end of its useful life. This removal cost could perhaps be as much as several hundred dollars per kilowatt, depending on the project's design and the site. Ask a solar service provider or general contractor for a budget cost to remove the facility. Remember to account for inflation since these costs will be incurred at the end of the useful life of the solar array. You may choose to wait to begin funding the reserve until you are closer to the end of the operational life of the system and have a better sense of when you are likely to need to remove it.

The underlying building may also need roof repair or replacement when the solar array is removed. Ensure that there is a plan to address the condition of the roof under the solar equipment at that time. In addition to routine annual roof inspections, perform a comprehensive roof survey periodically throughout the life of the solar facility. This will give you a good idea of the condition of the roof and will allow you to adjust your replacement schedule accordingly.

A major capital cost for solar projects is inverter replacement. While inverter technologies are quickly improving, many inverters are warranted for 10–15 years, whereas the solar modules are typically warranted for 25 years. A capital reserve can be included for this type of expense. Large one-time costs such as inverter replacement can have a big impact on overall financial performance, and may even exceed the revenue generated in that year. Aside from having a reserve fund, a second strategy to reduce the impact of large capital costs is to split it over two years to reduce the drain on cash flow in each year. While this is a benefit from an accounting standpoint and it does not change the cost of the replacement, it reduces the drain on cash flow in each year. This solution can be implemented by purchasing the inverter in one accounting year and performing the installation and related work in the following year. This may not always be possible if the replacement cannot be scheduled to occur at the end of the accounting year.

Table 8.5 Central inverter vs. string inverters

	Single "central" inverter	Multiple "string" inverters
Advantages	Cost-effective Space-efficient Extended warranty may be available	Replacement can be phased in over more than one year Replacement can occur without shutting down the entire project
Disadvantages	Full replacement cost typically incurred in a single year Requires full system shut-down	Higher initial cost May require a larger mounting area May have a shorter warranty period

Another solution that can help to manage large capital costs is to use a greater number of smaller-capacity inverters in place of one large inverter. When it comes time to replace the inverters, one or several small inverters can be replaced individually at lower cost over several years to spread out the expense. This also means that if one inverter were to fail it would only affect a portion of the solar array instead of the entire system. Designing the system for several smaller inverters may have a higher initial project cost, but the increased flexibility and redundancy could be valuable if and when it becomes time for replacement.

Project execution

Solar projects under the direct ownership structure can have a wide range of project delivery methods. A common route is to hire a solar development company to provide EPC services and turnkey project delivery. In other cases it may make sense to individually hire the services you need from engineers, contractors, and other vendors. This can be valuable when you are considering a specialized design, where solar is being integrated into a larger building development project, or where you simply want to exercise greater control over the project. In this case you will have more involvement as a project manager coordinating various consultants. In either scenario you will need an attorney to review contracts, and you will also want a tax advisor to weigh in on tax matters related to the proposed project.

The construction process must also be scrutinized to ensure it does not negatively impact your property, building tenants, or otherwise create unnecessary risks for your project. Staging materials on the site, construction traffic, and noise are some of the ways the solar project can affect the building and its tenants. Setting expectations upfront with tenants can alleviate many of these problems. It is also important to address this in construction contracts by specifying what the hours of operation will be, what areas are set aside for staging of materials, and where construction vehicles can be parked. You may also choose to hire qualified contractors that you have a good working history with. In addition to having the necessary experience on solar projects, a strong existing relationship can go a long way toward making a project proceed smoothly.

Operations and maintenance

O&M costs for a solar array tend to be limited since there are no moving parts and there are few components to physically wear out. Regular maintenance and up-keep is nonetheless required to keep the array operating in a safe and efficient manner. This will include costs for items such as:

- inspections;
- system performance monitoring and troubleshooting;
- cleaning solar modules;
- replacing modules due to non-warranted damage such as vandalism;
- repair or replacement of worn out electrical components.

Solar service providers and product manufacturers can help you budget for ongoing operational costs and future capital expenditures. Also consider the purchase of extended warranties for large capital equipment such as inverters or transformers. When you are ready to decommission the solar array and remove it from your property at the end of its operational life, you will be faced with the cost to remove and dispose of the modules, racking system, and all associated equipment. This may include removing and patching roof penetrations that were installed to anchor the solar array. Work with a roofer to develop a budget for roof-related expenses prior to the end of the solar array's life. In many cases replacing the roof in its entirety can minimize deferred roof maintenance costs. After all, by the time the solar array is removed the underlying roof will be at least 20 years old. If a roof replacement is not required, a roof repair budget should be identified one to two years before the array is to be removed.

Common variations on the direct purchase structure

This chapter described the basic form of the direct purchase structure in order to identify its key components, risks, and opportunities. The direct purchase structure can have several variations beyond simply funding a project out-of-pocket and hiring a solar company to provide turnkey project delivery. Alternative solutions you may want to explore include:

- creating a joint venture structure where two or more parties share investment costs;
- a capital equipment lease structure where you lease the equipment for a period of time, with a future purchase option;
- using short-term or long-term debt to fund part of the purchase;
- separating EPC services to have greater control over certain aspects of the project;
- integrating solar into plans for a building development or redevelopment project.

Other variations beyond those described above also exist. Depending on your needs you can work with your solar service providers to deliver the scope of services that best fit your objectives. For some property owners, this level of control is an important value and a reason to pursue the direct purchase structure.

Final thoughts

Property owners that have capital available to invest in a solar array can reap substantial financial and reputational benefits under the direct ownership structure. Financial benefits are seen through lower and more stable utility expenses, revenue from electricity sales to tenants, the capture of available incentives, and potential REC value. Additional value can be derived from the sustainable marketing aspects of the project, not only to tenants in the building but also to the local community and regulators. While you take on risk by investing in a solar array, you can choose how you want to manage that risk by controlling your level of involvement in the project and making decisions about the project's design and implementation. Chapter 9 explains the PPA, another common project structure that shifts risks away from the property owner and provides a lower-risk alternative to the direct purchase model.

Chapter 9

Power purchase agreement

Chapter summary

- The power purchase agreement (PPA) is a low capital cost project structure for deploying solar on commercial properties.
- PPAs require a creditworthy buyer of the solar electricity to sign a long-term energy purchase contract.
- Offsetting building energy use generates cost savings; selling solar electricity to building tenants generates revenue.
- The solar project developer retains the project's renewable energy certificates under most PPA agreements.
- Account for vacancy risk when selling solar electricity to building tenants.

The PPA is the second solar project structure that is commonly used for solar projects on commercial properties. The PPA structure represents a low capital cost solution for the property owner to deploy solar. Property owners stand to benefit from lower energy costs or revenue from re-selling energy to tenants, without incurring capital costs or other operational expenses. The value of the solar array for the property owner is derived from the value of electricity cost savings compared to utility energy prices, or from the revenue generated by selling electricity to tenants. The property owner has no ownership stake in the solar equipment under a PPA, so there is no claim to its value as a physical asset.

The PPA requires the property owner to purchase any electricity generated by the solar facility throughout the term of the contract, usually 20 years. The property owner can use the electricity to offset their own needs, sell the energy to building tenants, or some combination of both. The PPA offers certain benefits – such as stable solar electricity prices, peak energy cost-shaving, and marketing benefits – without the capital costs or ongoing operational requirements associated with direct ownership of a solar array. In the most common PPA scenario, a solar developer or EPC develops the project on its balance sheet after they have secured a PPA contract with a property owner to purchase electricity from the solar array. Once the solar array is operational, the property owner begins purchasing the electrical output from the project. The solar developer may subsequently transfer ownership to a third-party solar investor. The solar developer may remain involved in the project as the provider of operations and maintenance (O&M) services.

Revenue

The PPA structure can be set up to generate revenue for the property owner in several ways by:

* replacing a portion of grid-purchased electricity with lower cost solar energy;
* re-selling solar electricity to building tenants at a price greater than the PPA cost;
* reducing building peak energy demand charges.

The goal of a PPA is typically to provide solar electricity at a rate equal to or lower than today's total utility electricity costs. In this scenario the value created is the difference between the cost of utility-supplied electricity and the cost of solar electricity supplied under the PPA structure. Buildings with PPAs effectively have two electric utility providers – the traditional grid utility and the provider of solar energy. Providing solar can also serve as a marketing opportunity to attract building tenants that have an interest in being located at a property that supports clean energy.

Electricity value

The price at which the electricity is purchased can be set in several ways. One way is to establish a fixed price equal to today's utility electricity price. Over time as grid electricity prices rise the cost of solar electricity would remain constant. Another solution is to set the initial PPA rate at a price lower than the market rate for electricity, but to index it to inflation so that it is always a certain percentage below the then-market price for electricity. This discount is usually set at 10–20 percent, but can be set at any amount. There are other variations such as having a fixed cost that resets every few years or whenever a new tenant moves in.

Table 9.1 summarizes several solar PPA models to show the different way electricity costs can be determined when selling power to building tenants. Figure 9.1 provides a visual depiction of these various pricing scenarios compared to utility grid electricity prices. While these numbers are for indicative purposes only, they illustrate a range of potential solutions for determining what price to charge for solar electricity.

The solar project owner may also enter into their own PPA in order to sell power to one or more tenants. In this case the property owner will typically mark-up the cost of the solar electricity when they pass it along to the tenant. In some cases a tenant that wants to enter into a PPA directly with a solar company may approach the property owner. This would bypass the property owner in the PPA transaction. This may be acceptable for some property owners that are willing to support tenant solar efforts and do not want to insert themselves into the transaction. Other property owners may want

Table 9.1 PPA electricity cost comparison

Electricity source	Base price	Change in price
Utility grid	Market rate	Increases based on utility rate-setting
Solar PPA 1	Discount to market rate	Increases with utility prices
Solar PPA 2	Discount to market rate	Increases with pre-defined steps
Solar PPA 3	Above-market rate	Fixed throughout PPA term

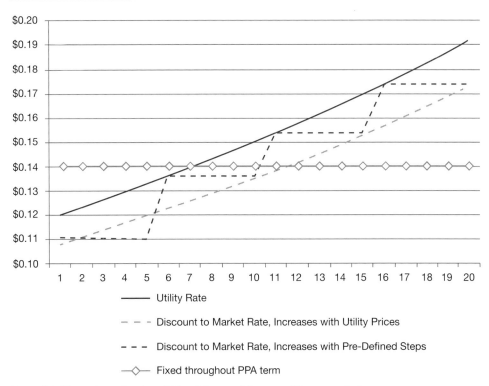

Figure 9.1 Illustration of various PPA pricing models over a 20-year period.

to receive some amount of roof rent or other payment for the use of the building's space for the tenant's solar PPA project. The property owner's ability to ask for this is usually a factor of the tenant's rights to use areas of the building to host solar in accordance with their lease. The choice is also a business decision about what is reasonable to seek as compensation for the use of space at the property.

Financial model

The financial model for the PPA structure is a simplified version of the one developed for the direct ownership structure in Chapter 8. There are no costs to the property owner for construction, O&M, or system removal at the end of the PPA term. The only likely out-of-pocket costs to account for are deal costs, such as legal fees to negotiate the PPA, roof inspections, and any repairs that may be agreed upon to prepare the building to host solar. Costs are offset by the value gained from reducing more costly grid electricity purchases or by re-selling energy to building tenants. As long as the value – in energy savings or in revenue – is sufficiently greater than the cost of the PPA, the solar project will create value for the property owner. Some property owners will want to maximize the revenue they gain from the project, while others will want to split the value and provide a portion of it to their tenants as a way to differentiate their property and increase tenant retention.

To construct the PPA financial model, include a row for the power purchase contract obligations. Add a row for the energy savings or energy sales revenue you expect to receive in each year. Account for taxes on the revenue you receive from the sale of electricity. Be sure you understand how the PPA payments you are responsible for are calculated in the PPA contract. The same considerations for adjustments to the price of electricity apply if you are re-selling electricity based on the terms of an agreement you have established with tenants. These terms are often aligned with the terms of the contract with the PPA provider in order to simplify administration of both contracts.

Renewable energy certificates

Under a PPA, the solar project developer may retain the renewable energy certificates (RECs) produced by the solar project. REC ownership can often be negotiated. Where there is a market for RECs that are needed to subsidize the solar project developer's costs, it is unlikely that the RECs can be provided to the owner unless they are purchased at market value or close to it. If this purchase is rolled into the PPA, it will have the effect of driving up the PPA price. In markets where RECs have little or no value there is likely to be more flexibility on the part of the solar developer to sell or assign the RECs to the property owner. But in that case the RECs are unlikely to have much value for the property owner either.

Risk profile of PPAs

The PPA structure eliminates many of the risks to the property owner that we saw in the direct ownership model, such as funding the construction, operating the solar facility, and capturing incentives. This includes the capture of tax benefits, which is one reason PPAs can be attractive for property owners that have limited tax liabilities. These risks are transferred to the solar developer that provides the PPA. These parties seek to manage their risks as well, and they will have several concerns that need to be addressed by a property owner before they are ready to sign a PPA contract.

Credit risk

The solar developer providing the PPA will need assurances that you are unlikely to default on the contract, leaving them without a buyer for their solar electricity. In a perfect world, property owners would have investment-grade credit ratings that minimize the risk of default for the PPA provider. In reality, few property owners aside from the largest companies can offer that level of publicly rated investment quality. This means that smaller property owners may have to spend more time and effort to meet the underwriting criteria to qualify for a PPA. Each PPA provider will have slightly different underwriting requirements.

Many properties are owned by single-purpose entities, such as limited partnerships and limited liability corporations. Some tenants may also operate under this structure. LLCs are not credit-worthy entities on their own. The parent entity of the LLC in most cases will need to be a party to the PPA in order to secure financing. Structuring the PPA contract to ensure that credit-worthy counterparties are involved can add a layer of complexity to PPA underwriting that may not be readily apparent when first discussing PPA opportunities with solar service providers. This is one of the reasons that the due diligence spreadsheet from Chapter 4 included a column to identify the type of ownership entity that held the property.

As the perceived risk associated with the power purchaser in the PPA contract increases, the value available to the property owner is likely to decline. This is because the PPA provider and their financial partners will want additional financial security against potential default. This could be less favorable PPA terms due to higher borrowing costs, or even a requirement that the power purchaser fund a default reserve. These provisions could be relaxed in the later years of the PPA as the solar investment is amortized. If you encounter this credit hurdle with one PPA provider, you may be able to find another PPA provider that is willing to take on the project. If you think this is going to be a potential issue given your property ownership structure or the size of your operations, discuss it up front with solar companies. This will help avoid proceeding with due diligence on a project only to find that the project cannot be financed with the PPA structure.

Vacancy risk

Property owners that seek to sell solar electricity to tenants in triple-net lease situations are exposed to what is known as vacancy risk. This is the risk that there may not be tenants available to purchase the energy that you seek to re-sell under the terms of the PPA. If you can't use the electricity for your own needs, you may face a loss of revenue. Selling power to multiple tenants can mitigate this risk; if one tenant departs, the others would still consume the power being generated. If this is combined with offsetting some or all of the common area energy costs, the net vacancy risk may be minimal. If this were done, only in cases of significant vacancy would this potentially become a problem.

To avoid this situation, some property owners only consider solar on properties where they are responsible for providing energy to tenants under gross leases. Other property owners focus only on solar to offset common area energy needs rather than the needs of tenants. When property owners are willing to sell power to tenants, the most attractive buildings are those that have stable long-term occupancy with leases to tenants with strong credit. While not common, ideally tenant lease terms would be comparable to the term of the PPA.

Property owners that choose to sell power to tenants should underwrite vacancy risk in line with the expected vacancy of tenant spaces. You may be able to negotiate to share the power sales risk with the solar company. This could be by sharing in the value of lost revenue or by capping the duration of your energy purchase obligation in the event of a protracted vacancy. These provisions may come at a cost in terms of less attractive PPA prices if the solar company believes they are sharing a meaningful risk.

An additional cost that is often overlooked is the time and expense required to sign up tenants to purchase solar energy. Negotiating a power sales contract with current tenants and any future tenants can add time and uncertainty to the PPA process. A solution to address this is to have the power sales agreement incorporated into the standard lease at the property, so it is part of the deal from the outset rather than being added on separately. While tenants receive a benefit from the solar array, the effort required to incorporate that benefit successfully into a lease should not be overlooked.

Quick tip

When selling electricity to building tenants, account for periods of tenant vacancy in your energy sales projections as you would with rental income.

Sale of property

Consider whether your PPA obligations could affect the future sale of your property. This risk could be triggered if there are requirements in the PPA that the power purchaser maintain a minimum credit rating. This may be an acceptable commitment for a large institutional buyer to take on, but a sale to a smaller owner-user with lower credit could result in default on the terms of the PPA if they cannot meet the credit requirements in the PPA contract. In order to complete the sale, the property owner could be forced to pay a penalty or a termination fee to get out of the PPA, or in a worst-case scenario, seek another buyer. Neither solution will have much appeal for property owners ready to sell their property.

Bankruptcy

The bankruptcy of any party related to the PPA can affect the ability of the solar project owner to meet its obligations of providing electricity to the property owner. These parties could be the solar project developer, system operator, or financing partner. This warrants consideration because of the 20-year lifespan of solar projects. Over this length of time, history tells us that both the real estate and solar industries are likely to see one, if not more, economic cycles. A carefully constructed PPA contract and extensive due diligence will help to manage these inevitable periods when industries could face systemic risk in a slow economy. Agreements should have provisions to hold current or future project owners, whether they are a solar company, investor, lender, or another party, to the same lease obligations as the original PPA contract signatory.

A bankruptcy at the solar project could jeopardize the property owner's revenue from energy sales, RECs, or energy savings value. While unlikely, if the bankruptcy results in the solar project being abandoned, the property owner could also face the cost of removing equipment from their building. Termination rights in the PPA contract should provide an exit path in order to protect the property owner from a non-performing project due to a bankruptcy or other default. Another solution could be a buy-out provision in the contract that allows the property owner to take over the project. This would enable the property owner to benefit from the solar array if the original owner

Quick tip

Protect your project from bankruptcy of the solar developer by including termination rights in the PPA contract.

cannot honor their commitments. Buy-out provisions may be based on a predetermined schedule of values, a fair market value appraisal, or they may be set as a function of the value of the energy produced.

Existing lenders

Debt carried on a property may affect your ability to successfully undertake a solar project. Lenders may have rights to approve any new financial obligations that could affect the ability of the property to service the lenders' debt. Signing a PPA that obligates you to purchase electricity from the solar project, regardless of whether you have a tenant to re-sell the power to, may cause the lender concern. If the mortgage lender takes possession of the property in the event of a default, they could also step into the property owner's shoes for the PPA obligations. Subordinating the power purchase obligation to the mortgage is a solution to this concern, but the solar company that controls the PPA – and their lenders – will not like the subordination because it adds risk to their financial position. The specter of additional roof maintenance costs or other O&M obligations may also be a secondary concern for some lenders, although these concerns can be alleviated with a properly structured agreement between the property owner and the solar project owner.

It is important to highlight that lenders have the opportunity to benefit from solar. Additional revenue from a solar project can increase operating revenue and improve debt service coverage ratios. Solar project revenue is also diversified from the volatility of the real estate market. This is particularly true when there is a roof lease backed by a long-term power sales contract to a third party such as a utility company. A PPA at a gross leased property that is supplying power to offset landlord energy costs can also provide a hedge against rising energy prices that could erode the property's net operating income. Solar projects installed on properties that lenders have invested in also provide an opportunity for the lender to show its support for "greening" their investment portfolio. While this is unlikely to affect underwriting at the property level, it does provide a silver lining to their support for solar projects.

Cost considerations

Project structures such as PPAs are often pitched as having the advantage of "no cost" to the property owner. This may be true from the standpoint of the capital costs to develop the solar array and the obligation to maintain and operate it, but there will be costs for due diligence and contract negotiation on the part of the property owner. These may include:

- legal fees;
- roof inspections;
- proactive roof repairs;
- engineering peer reviews;
- tax/accounting review;
- brokerage fees;
- travel and other miscellaneous expenses;
- staff time.

One way to reduce these costs is to establish an upfront option fee paid by the solar company to secure the rights to the use of your properties. This fee may be set at a level sufficient to offset some or all of the out-of-pocket due diligence and negotiation costs. For one-off projects, this may be difficult for solar companies to agree to, but if your portfolio has the potential for deploying multiple projects, this is a more likely scenario. Another solution to recoup some of these costs is to include construction phase payments to compensate for your in-house project management time. These fairly modest fees can help you offset out-of-pocket planning and due diligence costs. Keep in mind, however, that shifting large costs on to the PPA provider could result in less attractive PPA terms.

You may find that due diligence costs add up quickly. As a result, there may be a temptation to cut corners on these out-of-pocket expenses, particularly when the solar company offers to share its engineering and roof conditions report with you. Conducting your own due diligence and engineering has several benefits. Aside from eliminating conflicts of interest, it helps to have a few extra pairs of expert eyes examining the proposed solar project to make sure preventable problems are not overlooked. When talking to lenders, joint venture partners, and investors, a thorough due diligence process will help you make a compelling case that the project will add value.

Project execution

Project execution for the property owner pursuing a PPA is fairly straightforward. Most of the time and effort for the property owner is spent on due diligence to find a suitable property and PPA provider, and then negotiating the terms of the PPA. Once the PPA is executed, the solar project developer will complete a detailed design, procure the solar equipment, and engage contractors to construct the array. Coordination is required to ensure that the project does not cause problems with the operation of the building or its tenants, but the bulk of the project is handled in a turnkey manner by the solar developer and any associated sub-contractors.

In advance of the solar project installation, you may be required to make repairs or upgrades to the building's roof so its lifespan will be comparable to that of the PPA agreement. Failure to do so could result in a large cost to the solar project owner to remove their array temporarily while you repair or replace the roof. Where this is a possibility, particularly on older rooftops, most PPA providers will insist on protections to mitigate their financial risk. The most common remedy is proactive repairs, coatings, or roof replacement. If risk remains that could lead to a large future cost due to roof work, it will likely translate into less attractive PPA terms or an obligation that the property owner shares in the removal and reinstallation cost. Alternatively, you may choose to take on the entire risk and agree to pay for any temporary solar relocation costs and downtime if extensive roof repairs become necessary.

Quick tip

Shifting risks and costs to the solar project developer has trade-offs; it reduces the attractiveness of the terms offered to you in the PPA.

Operations and maintenance

Under the PPA structure, the owner of the solar array is responsible for all ongoing operational costs, maintenance, and upkeep. Unlike the direct ownership structure, the property owner hosting the solar project does not bear these costs. The solar project owner will hire a company to perform ongoing system monitoring and upkeep. In some instances this is the company that constructed the array, but there are also companies that specialize in O&M for solar projects.

One of the most important areas of solar O&M to focus on is where the solar array interfaces with the host building. This is typically where the solar racking system contacts the roof, but similar recommendations apply if the array is part of a façade or other building element. The PPA should spell out which party is responsible for maintenance and repair of the roof, the flashings at solar equipment, support structures, and any other elements that are part of the roof and solar interface. It is important for this responsibility to be defined clearly for both parties to avoid finger pointing in the event a repair is needed. There are several ways to assign roof maintenance obligations in a PPA:

- The property owner may retain roof maintenance responsibilities, subject to the cooperation of the solar array owner. This may include temporary removal and re-installation of sections of the array if needed, or powering down portions of the array to permit safe access for roof repairs. This scenario allows the property owner to retain control of the roof – and the ability to remedy leak problems or other issues before they cause problems with other building tenants or damage the property. This also enables the property owner to use preferred contractors and ensure that the roof warranty is maintained.

- The solar array owner can take on the responsibility for maintenance of the roof. For projects that fill the roof of a building, this could be a reasonable expectation since almost any roof work would have the potential to impact their solar array. In the event that there are increased maintenance needs because of the solar equipment, the solar project owner would have responsibility for repairs, not the property owner. In this situation, the property owner may want to retain the right to require the solar array owner to use a preferred roofing contractor in order to ensure the quality of the repairs and maintain the roof warranty. The property owner may also specify that its roof inspection company be present during repairs or provide inspection reports following any work that is completed.

- Where the solar array covers only a portion of the roof, the PPA provider may assume responsibility for roof maintenance under the area of the solar array while the property owner retains responsibility for other areas of the roof. This is an equitable way to share responsibility for roof maintenance and repairs. In this scenario, coordination is required between the property owner and solar project owner to complete repairs. Both parties should agree on the contractor used for repairs to ensure consistent, warrantable work. Coordination will also be required in the event there is a roof repair needed in an area that overlaps with the zones each party has responsibility for.

The contract should specify what would be done to remove the solar array at the end of the PPA term and, if necessary, restore the roof to acceptable condition considering its age at that point in time. There are several paths that can be pursued at the end of the PPA term:

- The PPA could be extended for one or more years.
- The solar project owner could remove the solar array at their cost.
- The property owners could take over ownership of the solar array.

Removing the solar array is the standard solution at the conclusion of the PPA term. Extending the PPA term would be subject to mutual agreement that the array and the roof are in suitable condition and there is value for both parties to continue the relationship. A scenario where the property owner takes over the solar array is much less common. This scenario may make sense if the term of the PPA is less than 20 years and there is a fair amount of operable life remaining in the solar array. This can provide an opportunity for the property owner to squeeze additional revenue out of the system now that they have gotten comfortable with the operation of a solar project. Before pursuing this path, however, determine whether the incremental value of taking ownership of the solar array and its operations is worth taking on the potential operational costs and eventual disposal of the equipment. If you are considering this option, have the solar operator provide you with the maintenance records and cost history so you have an accurate picture of their run-rate on costs. This will also help you identify if there has been any deferred maintenance that could lead to problems or costly repairs in the future.

Common variations on the PPA

The basic framework of the PPA is flexible and many risks can be allocated between the property owner and the solar developer in accordance with the needs of specific projects. This provides a great deal of customizability that can meet the requirements of many property owners. Specific terms can be negotiated based on the project in question, including:

- the length of the PPA term – typically from 15 years to 25 years;
- project costs and risks can be shifted or shared between parties;
- marketing solar electricity can be the joint responsibility of both the property owner and the PPA provider, or it can be assigned to one party;
- roof maintenance can be assigned variously to either party or both parties in the transaction;
- roof repair or replacement costs could be incorporated into the PPA price;
- at the end of the PPA term, the project may be removed, the contract could be extended, or ownership may change hands.

Tenant-initiated PPAs

Property owners may be asked by tenants to support a tenant-initiated PPA project, where a tenant seeks to offset its own electricity needs with solar energy. In this scenario, the property owner may simply be asked to consent to the agreement. This is more likely to occur where the tenant in the building has a long lease term – 15 or more years – and where they see value in offsetting their on-site energy needs. Anchor tenants

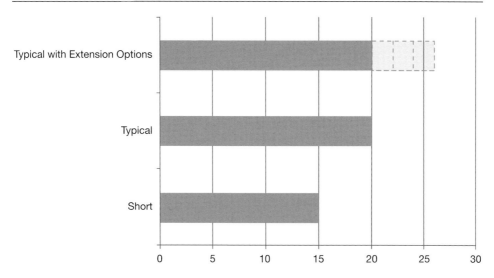

Figure 9.2 Range of typical PPA contract terms (in years).

at retail properties are one of the most common examples of this situation. These tenants generally have long-term leases and consistently high energy needs. They also often have building and roof maintenance responsibilities included as part of their lease. This minimizes the risk to the property owner asked to approve the project because they have few obligations, if any, with respect to the PPA. Just as importantly, for the solar project developer providing the PPA, they are able to sign an agreement with a retailer whose creditworthiness is likely to be easier to underwrite than the property owner.

With a tenant-initiated solar PPA, the property owner may seek to negotiate a roof lease with the PPA provider, subject to the terms of the existing tenant's lease. This is done to ensure that there is a contractual relationship between the solar company on the property owner's building and the property owner, and to enable the property owner to receive a modest amount of revenue for the use of that space. If the property owner would be giving up the ability to use the roof space for his or her own purposes, the property owner may be able to negotiate to receive roof lease payments as a condition of approving the project.

Final thoughts

The PPA structure offers a significant reduction in risks associated with developing and obtaining the benefits of a solar project. The use of a solar developer to finance, construct, and operate the solar project makes this solution attractive for many property owners that do not have the desire to become solar experts themselves. Other property owners are likely to find that the PPA is the most suitable solution because of their limited ability to effectively capture tax credits or other incentives. Chapter 10 reviews the third and final solar project structure: the lease structure. This provides an alternative to the PPA model for property owners that are interested in adding solar to their building, but have operational constraints that limit their ability to use solar electricity at the building or sell energy to tenants.

Chapter 10

Lease

Chapter summary

- In a lease structure, a solar developer leases roof or ground space at your property to host a solar array that they fund, construct, own, and operate.
- Solar electricity from a lease is sold to the utility company or an off-site third party.
- A lease allows property owners to generate revenue from hosting solar even if they are unable to offset common area energy use or re-sell solar electricity to tenants.
- Lease payments can be calculated based on fixed criteria such as the area occupied by the solar array, or they can be calculated based on variable measures such as energy production.

The lease structure is the third method for pursuing solar projects on commercial buildings. Under the lease structure a solar developer funds, constructs, owns, and operates the solar array in exchange for paying you to use space on your property. A third party purchases the electricity from the solar project developer under a PPA contract. This third party is usually a utility company, but in some markets it could also be another property or business. In contrast, recall that the direct ownership and PPA structures utilize electricity on-site to meet building energy needs.

In the lease model, the property owner does not put any of their capital at risk and has no power purchase obligations. The lease is most often applied to space on the roof of a building, although it could be applied to a canopy above a parking area or where there is other available space on the site. The subject of this chapter applies to any of these cases, but for simplicity we will refer to this type of lease as a "roof lease" to avoid confusion, and to differentiate this structure from a traditional tenant's lease for space inside the building.

The lease structure was popularized to enable solar development on properties that contain a desirable resource such as a large rooftop area or open ground space, but cannot readily use solar energy on-site due to operational or contractual limitations. This could be the case at a building that has triple-net leases where tenants are paying their own utilities directly without the involvement of the property owner. This could also be the case at a building that has a roof area large enough to support solar production well in excess of an amount that could be used on-site. An example of this is a large distribution center or warehouse that has a rooftop capable of supporting several megawatts of solar capacity, even though on-site energy needs are typically fairly low. Even property owners that could use the electricity on-site may choose to simply lease space. They may do this in order to increase property revenue in a way that is familiar – leasing space – and without adding to their administrative requirements, since there is no need to re-sell power to tenants or otherwise enter into the obligations of a PPA. A property ownership entity that is interested in a PPA but is not eligible due to limited creditworthiness may find the lease structure attractive as an alternative way to generate revenue and support clean solar energy.

Quick tip

Distributed generation refers to the production of electricity from numerous small-scale or mid-scale sources within a utility territory, often close to end users of electricity. Compare this with traditional central plants that generate large amounts of electricity far from the end consumer.

The lease model is currently available in a limited number of markets and is typically enabled by regulations that mandate electric utilities to purchase solar electricity from numerous "distributed generation" sources of renewable energy. To date, roof leases have tended to be targeted toward larger projects that can deliver sizable quantities of power to the utility grid compared to typical solar projects that only generate enough power for a portion of on-site needs. For example, in southern California, a utility distributed generation program supports leases and targets projects of 1,000 kW or greater in size per site. A number of the roof leases in that market have developed systems of 3,000 kW or more.

Lessons from the field

The lease structure vs. the "solar lease"

It is easy to confuse the lease structure discussed in this chapter with a popular residential solar financial product known as a "solar lease." For our purposes on commercial buildings, a lease describes a project structure that property owners engage in to generate revenue from their rooftop space without investing in the solar project or consuming the electricity it produces. A "solar lease" is a financing solution popularized in the residential solar market that enables homeowners to lease solar equipment on their house for a fixed payment. For commercial projects, leasing solar equipment in this manner is considered a variation of the direct ownership structure, which was covered in Chapter 8.

Revenue

The income generated by a roof lease is the primary source of revenue for the property owner under the lease structure. Payments are typically made monthly, but they may be made quarterly or semi-annually if desired. Roof lease payments can be either fixed or variable. Fixed payments offer consistent revenue that can easily be calculated. Variable payments can better align the interests of the property owner with the solar array operator, but may require greater auditing and oversight to ensure accurate payments. There are three common ways to calculate fixed lease payments:

1   Lease payments can be based on the gross square footage of the roof that is occupied by the solar installation. This area is calculated by drawing an imaginary line around the perimeter of the solar array to determine the square footage inside that perimeter. The roof lease payment is then calculated by multiplying that roof area by the roof lease rate. This calculation is often used when the solar installation occupies only a portion of the total roof area.
2   Fixed roof lease payments can also be based on the total gross roof area of the building. This is more likely when the solar array occupies the entire rooftop. The calculation of the roof lease payment in this case would tend to result in a lower per square foot rent, even though the total payment might be the same as when calculating the rent based on the area of the solar array only.
3   A third method for calculating roof lease rates is to base it on the rated power capacity of the solar array, measured in kilowatts. This more directly ties payments to the production capability of the array instead of the space it occupies. To calculate this lease rate, multiply the total rated kilowatt capacity of the solar array by the roof lease rate. Remember that each solar module's rated capacity is measured in watts, so you will have to divide by 1,000 to determine the number of kilowatts; a module rated as 300 watts is equivalent to 0.3 kilowatts.

If you decide to base the lease payment on the roof area, recognize that it may be possible for the solar project operator to add additional solar generating capacity in the future without needing to lease additional space from you, since they already have rights to space on the roof. They could do this by replacing outdated modules with newer high-efficiency modules in the future. While this is not likely, it is possible. To protect against this possibility you may want to include a provision in the lease that requires your sign-off for any changes to the solar project that affect the overall size or capacity of the array. The alternative, basing lease payments on the installed capacity of the solar array, allows you to modify the payment if the system capacity increases.

Lease payments can also be based on variable measures of solar project performance. Variable measures could include the kilowatt-hour output of the array or a percentage of the revenue generated by the power sales contract the project owner has with the utility. These methods of calculating roof lease revenue will vary based on the output of the solar array due to weather and other operational factors that are outside the control of the property owner. This type of payment does tend to align the interests of the property owner and the solar operator, although the property owner has little ability to affect solar array performance. For property owners that do not want to miss out on any over-performance by the solar array or potential increases in value from the sale of

> **Lessons from the field**
>
> *Payments in advance or in arrears*
>
> Pay attention to whether lease payments are to be made in advance or in arrears each month. Paying in arrears means the payment occurs after the time period has concluded, much like you would pay your phone bill based on your prior month's usage. Payments in advance mean the payment covers the upcoming time period, as you would do with a mortgage payment. Since traditional tenant space leases specify that payments be made in advance and real estate companies are set up for this payment schedule, it is usually easiest to use this format when leasing space for solar. The solar project owner may prefer payments in arrears, however, in order to better manage project cash flow. This enables them to be paid for the array's output, and then pay for the use of space out of that payment.

solar electricity, this type of lease payment calculation may be beneficial. Conversely, property owners such as real estate investment trusts may want to avoid variable lease revenue because it can be classified as non-qualified or "bad" income for purposes of maintaining REIT tax status.

Financial model

Creating a financial model for the lease of roof space for a solar array is similar to the model used to evaluate other leases at your property. You receive income in exchange for the use of space at your building. The solar operator is responsible for operating expenses to maintain the solar array so you do not need to factor these in. Understand how the lease calculates rental rate increases – whether it is a fixed increase, based on an inflation index, or some other measure. Include whether the lease payments occur in advance or in arrears since this will impact when revenue is received. If the roof lease includes variable payments based on system performance or electricity sales, be sure to account for the likelihood of system downtime, changes in energy prices, and the variability in lease revenue that could occur throughout the year. Account for any taxes that may need to be paid on solar revenue.

Renewable energy certificates

Under a roof lease, the solar project owner controls the renewable energy certificates (RECs). If the solar operator is selling the project's electricity to a utility, the utility will take title to the clean energy attributes of the electricity as a condition of contracting with the project operator. As a result, the property owner does not have a claim to the RECs.

The property owner should also be careful about describing their participation in the solar project to avoid the possibility of misinterpretation by investors, customers, and the general public. Because neither the RECs nor the electricity are controlled by the property owner, it is inaccurate to imply or make claims about having a "solar building" or even "creating clean energy." These comments could suggest a claim to the RECs that are rightfully owned by another company. It is most appropriate to limit comments to the extent that "the building leases space to a third-party owned solar array." This will help avoid any confusion about ownership of the clean energy attributes from the solar array.

Risk profile of leasing

The lease structure is the lowest financial and operational risk solar deployment structure available to commercial property owners. The property owner has no financial investment in the solar project, as they would with the direct purchase structure. They have no obligation to purchase energy from the array, nor do they have to find a buyer for the electricity as would be required with a PPA structure. The lease structure simply adds a long-term tenant to the building. The ongoing management of this rooftop tenant can be handled by existing property management staff. Unlike many other building tenants, most rooftop solar projects require little ongoing oversight from the property manager once they are operational.

Cost considerations

Out-of-pocket costs for the property owner under the lease structure are limited. As you saw with the PPA structure in Chapter 9, the property owner will incur costs to negotiate the roof lease and perform inspections. Upfront investments for due diligence and to prepare the building for solar will be money well spent, particularly since solar projects under the lease structure tend to be large and cover a substantial portion of the building. These efforts also eliminate potential conflicts of interest and avoid over-reliance on information provided by the solar developer. Costs for the lease structure may include:

*   legal fees;
*   roof inspections;
*   proactive roof repairs;
*   engineering peer reviews;
*   staff time.

During the design and construction of the solar array, coordination is required to ensure that the project complies with structural limitations for the building and that the building condition is suitable to support the solar project. You may want to have your own structural engineer perform a peer review of the solar company's structural calculations. If these obligations are managed properly they set the stage for a low-risk lease that generates stable long-term income from under-utilized space on the building.

Project execution

Project execution for the property owner pursuing a lease is fairly straightforward. Most of the time and effort for the property owner is spent on due diligence to find a suitable property and negotiate the lease. Once the lease is executed, the solar project developer will complete a detailed design, apply for and receive interconnection approval, procure the solar equipment and engage contractors to construct the array. Coordination is required to ensure that the project interfaces smoothly with the operation of the building and its tenants, but the bulk of the project is handled in a turnkey manner by the solar developer and their contractors. Compared to the PPA structure, lease projects tend to be larger. As a result, the planning and construction are likely to last longer, increasing the potential for delays or tenant disruption. Conversely, most lease-based projects interconnect to the utility grid at a point separate from the building's electrical service. This eliminates a potential impact to the building since it is not necessary to shut down building electricity supply to interconnect the solar array.

In advance of the solar project installation, you may be required to make repairs or upgrades to the building's roof so it is capable of lasting for the duration of the lease agreement. In some instances it may be possible to incorporate the cost of replacing an aging roof into the terms of the lease. This will result in a lower lease payment but it would eliminate a costly building capital expense.

Quick tip

Ensure that the solar project owner carries insurance for their property and operations comparable to what you require from other building tenants.

Operations and maintenance

Under the lease structure, the owner of the solar array is responsible for all ongoing operational costs, maintenance, and upkeep. The property owner hosting the project does not have responsibility for these costs. The solar project owner will typically hire a company to perform ongoing system monitoring and upkeep.

One of the most important aspects of solar O&M that can affect the property owner deals with the locations where the solar array interfaces with the host building, typically the roof. The lease should spell out which party is responsible for maintenance and repair of the roof-to-solar interface. There is no right or wrong way to assign roof maintenance responsibility, as long as the responsibility is clear for both parties in the event a repair is needed. There are several ways to assign roof maintenance obligations in a roof lease:

- The property owner may keep roof maintenance responsibilities, subject to the cooperation of the solar array owner. This scenario allows the property owner to retain control of the roof – and the ability to remedy leak problems or other issues before they cause problems with building tenants or lead to extensive damage. This also enables the property owner to use their own contractors and maintain roof warranties.
- The solar array owner could be made fully responsible for maintenance of the roof. This may make sense if the project covers the entire roof of a building. The property

owner may want to retain the right to require the solar array owner to use a preferred roofing contractor and follow pre-established roof inspection procedures.

- For projects where the solar array covers only a portion of the roof, the solar project owner may take on the responsibility for roof maintenance that falls within the perimeter of the solar array. At a minimum, both parties should share the contractor used for repairs to ensure consistent, warrantable work.

At the end of the roof lease term, the solar array owner should remove the solar array at their cost unless a lease extension has been executed. This is comparable to the obligations of any other building tenant that is required to remove their furnishings and other equipment at the end of their lease.

Common variations on the lease structure

The most common variation on the lease structure is a hybrid structure that combines the roof lease with the PPA. Recall in the PPA structure when a tenant requests that the property owner approve a tenant-initiated solar PPA project. In this case the property owner would execute a lease agreement in parallel with the tenant's execution of the PPA. From the tenant's perspective they have entered into a PPA. From the property owner's perspective they have entered into a roof lease. This situation is likely to occur when the property owner controls the roof or ground space where the tenant wants to install their solar array. This is a hybrid structure for the property owner since it incorporates a lease, even though building tenants use the solar electricity on-site.

Final thoughts

The lease structure enables unused space on a roof or on the site to become an income-producing asset. Leasing space on your property provides a low-risk way for property owners to benefit from the revenue-generating opportunity that solar represents, without the need to invest in owning a solar project or committing to purchase electricity under a long-term contract. Compared to the direct ownership structure, the lease structure has a significantly reduced risk profile by eliminating property owner responsibility for funding, developing, and operating a solar project. The lease structure shares similarities with the PPA model in that there is limited risk to the property owner for project finance and project development. Where they differ is in the significant long-term power purchase commitment required for the PPA compared to a fairly straightforward contract to lease space with the lease structure.

Each of the three project structures presents a distinct profile for risks and potential returns for the property owner. For those that have sufficient resources, a tolerance for risk, and a goal to maximize revenue, the direct ownership structure may be attractive. For others with good credit that seek the benefits of using or selling solar power at their property, a PPA may be a suitable solution. For those seeking incremental revenue while minimizing their day-to-day involvement in the solar project, the lease structure may be the best fit.

Other project structures exist and it would be difficult to identify them all. You could, for example, create a business model to develop solar projects in-house with a goal to sell power to the utility or another third party. In this case you would be using the direct ownership model along with the lease structure. While not yet common, you could – if regulations permit – enter into a PPA for solar electricity from a solar array located off-site at another property. In this case you would take on the benefits and obligations of the PPA without needing to deal with the impacts of the solar development process on your property. These are a few of the many other scenarios that are less common, but still possible given the right conditions.

The information contained in Part III described the knowledge you will need in order to choose the structure most suitable for your goals and risk appetite. If your needs do not fit neatly into the direct ownership, PPA, or lease structures, you should be able to identify the aspects of each one that may enable you to create a hybrid structure to meet your goals. Next you will move to Part IV and begin a discussion of potential risks and how to identify and mitigate them as you pursue solar projects.

Part IV

Identifying and managing risks

Solar modules have been in commercial use for more than half a century, but it is only in the last decade that solar has seen widespread adoption for commercial property use. As the solar industry has scaled up, technologies have evolved. The ranks of solar installers have swelled. Capital has emerged to fund projects and develop new operating models that support the widespread deployment of solar for commercial properties. The rapid evolution of the solar industry has created new opportunities to deploy solar profitably, but it has also contributed to certain risks that can impact the success and profitability of projects. These risks can be broadly categorized in two groups:

1   physical risks;
2   systemic risks.

Taking the steps necessary to deliver a successful project means that some of these risks will inevitably arise. To effectively manage them and mitigate their potential impact, property owners need to understand what they are and how to handle them. Some of these risks arise when planning the project. Others emerge during the construction process. Certain risks appear when it is time to remove the solar array from the building at the end of its useful life.

Thousands and thousands of solar projects are built each year, and over their lifetimes the projects that create the greatest value for property owners will be those best able to understand and manage risks. Part IV of this handbook provides the tools to ensure your project is one of them. The information contained in the following chapters provides a comprehensive view into the risk factors you need to understand as you plan and execute your solar project. Identifying the most likely risks empowers you to weigh trade-offs, seek actionable solutions, and implement them in a way that ensures the success of your solar project.

The risks identified and explained in Chapters 11 through 15 are primarily under the control of the property owner and solar project developer. These are predominantly physical risks, and they can be anticipated and managed effectively through proper due diligence, engineering, and construction. While there may be decisions to make and costs involved, a solution lies largely within the control of the project team.

The second group of risk factors is non-physical in nature. This category, the focus of Chapters 16 and 17, represents potential project impacts that are outside the direct control of the property owner, solar developer, and project operator. These systemic risks have the potential to impact not just your project, but also many projects or the entire industry. Changing regulations in both the solar and real estate industries could affect – either positively or negatively – the value of current and future solar projects.

When it comes to deploying solar on commercial buildings, the ability to anticipate and mitigate risks is one of the most important skill sets needed to deliver projects that are accretive assets for the underlying property. Successful risk management is one of the hallmarks of a solar program that consistently generates revenue and makes the underlying property more valuable.

Chapter 11

Permitting, design, and engineering

Chapter summary

- The permitting process can be lengthy; build in ample time to avoid delays.
- Ensure there is sufficient space around existing rooftop equipment and adjacent tall objects to avoid shading the solar array.
- Coordinate the design of the building and the solar array to maximize the space available for solar modules.
- Perform a structural analysis to ensure that both the racking system and the building structure can resist the loads imposed by the solar modules.
- The ability of a building to support the weight of solar equipment may dictate the choice of racking system and affect the size of the solar array.

Permits and code compliance

Issuing construction permits for solar facilities is a fairly new process for many building inspection departments, and officials may not have much familiarity with the design and operational characteristics of solar arrays on commercial building rooftops. Overlooking the impact on schedule that the permit process has could leave your project exposed to the risk of delays. The better the local code officials understand your project the more likely you are to receive approvals in a timely manner. You can avoid this pitfall by ensuring that either you or the solar developer has checked local codes. Where possible, review plans with local code enforcement officials early in the design process to identify questions that need to be addressed.

Additional reviews and approvals

The solar permitting process may include reviews by entities that may not be involved in traditional construction projects. Local fire department officials may have specific requirements that are not spelled out in the building code. They may have recommendations for paths of travel and access to the roof that need to be factored into the design of the solar facility. They may also have specific guidelines for marking safe paths of travel on the roof for fire department use that will need to be incorporated into the project. While not usually a big cost, the project should seek this feedback and incorporate it in order to avoid any last-minute delays or redesign.

For projects that interconnect to the power grid, the utility will also review the project to assess any impacts on their electrical distribution grid. In some solar markets the utility review process could be very brief, while in others the process requires significant paperwork and months of time. Conversations with the utility will often begin during the system design phase, but may not be fully complete until the project is ready to start construction. Consult with a solar developer to get a sense of how this is handled in your market.

Zoning review

Solar projects may require review by the local zoning board before proceeding with construction. The approval process might also include historic commission sign-off, depending on the placement of the solar array and whether or not it will be visible from the street. Solar equipment that is permanently attached to the building – as opposed to ballasted to the roof – may be considered an integral part of the building from a zoning standpoint. This could require compliance with zoning height restrictions or setback requirements that may be less stringent for equipment that is not permanently attached. The same is also true of solar facilities that are constructed as awnings or canopies for parking lots. Here again, it can be beneficial to be proactive and seek input about your project early in the design stage to identify any requirements the zoning board may have and any flexibility that may exist. This is particularly important if they have not reviewed a solar project like yours in the past. Concerns may be avoided by designing the solar project to lie flat or to sit below parapet walls so it is not visible.

System design and layout

The design of the solar facility has a significant impact on the cost, quality, and performance of the project. Decisions made during the design phase of the project can affect:

- cost;
- energy output;
- reliability;
- serviceability;
- operational flexibility;
- aesthetics.

Clearance for existing equipment

The solar design engineer utilizes as-built drawings of the property to determine the size and layout of the solar array. If as-built drawings are not available the solar developer may commission a survey to map out the target area. The designer will also review nearby buildings or large objects such as billboards or trees that could cast shadows on the solar array, making parts of the property less suitable for solar modules. The designer also needs to know if you are planning to reserve a portion of the solar site for future equipment or for tenant use, or if use of certain areas is restricted to certain sections. Ensure that the design maintains sufficient access to all sides of rooftop equipment, such as:

Figure 11.1 Carport with a solar array for an overhead canopy (Source: photo courtesy of US Navy, via Wikimedia Commons).

Note: Each of the two canopies has a roof composed of solar modules. The carports provide shade as well as clean electricity for the site.

- heating, cooling and ventilation equipment;
- roof hatches and ladders;
- elevator penthouses;
- skylights and clerestory windows;
- roof curbs and flashings;
- roof drains and scuppers;
- low points on the roof where ponding could or does occur;
- satellite dishes and antennas;
- gas pipes;
- expansion joints.

You may find that your property has few obstructions or clearance restrictions and therefore supports a solar project large enough to cover a significant portion of the building. This would allow you to maximize revenue from the array by making it as large as possible. It would, however, restrict your ability to inspect and maintain the roof. This makes equipment maintenance and replacement more difficult and therefore more costly. It can also have dangerous consequences if adequate safe work areas are not provided near roof edges or skylights. Ensure that there are setbacks around all rooftop equipment, skylights, flashing, and parapets so they can be inspected and repaired without compromising the efficiency or safety of the repair work.

Figure 11.2 Solar module obstructing access to a metal vent hood.

Note: This metal vent is partially covered by a solar module. Over time, moist and hot exhaust air from this vent could damage the solar module. This vent also cannot easily be inspected or maintained without first removing the solar modules.

Figure 11.3 Solar array clearance at a rooftop fan.

Note: A solar panel was left out of the array behind the tall metal fan (far right of the photo). This prevents the fan from casting shadows on the array. A white combiner box was installed in the resulting open area – an efficient use of space when a solar module cannot be installed.

During the solar survey also verify that nearby objects that could cast shadows on the roof and degrade the performance of the solar facility are documented. Obstructions adjacent to the site could include:

• buildings;
• trees;
• cell phone towers;
• billboards, signage, and flagpoles;
• highway ramps and overpasses;
• power lines.

The survey should also note features of the site and zoning that could affect the placement of electrical equipment such as transformers and inverters that will be located on the site. These could include:

• driveways and parking areas;
• fire lanes;
• zoning setbacks;
• easements;
• underground gas, water, and sewer pipelines;
• underground vaults, tanks, and electrical duct banks;
• subsurface monitoring wells;

- root systems of trees;
- sight lines for building signage;
- expansion areas reserved for building tenants.

Accounting for existing conditions from the early stages of design reduces the need to make subsequent changes to the solar array layout. This also provides a quicker and more realistic assessment of the potential size of the solar array, and enables more accurate budgeting. For example, ignoring setbacks around the roof perimeter until well into the project's design could result in a last-minute reduction in project size that affects the overall economic value of the project. The more clarity you have about the project's size and scope, the less risk there will be late-stage changes that impact project feasibility.

Incorporating solar into new building development

Planning to construct a building where solar is incorporated into the design is an effective way to ensure that the building and the solar array are fully compatible with each other. As was noted in Chapter 6, coordinating the requirements of the building design and the solar array can reduce project costs, eliminate conflicts, and improve the longevity of the solar array. Where feasible, have the design team's engineers co-locate rooftop equipment in areas that minimize their impact on the layout of the solar facility. For example, place a tall cooling tower on the north side of the building so the shadows it casts do not reduce the area suitable for solar modules; align skylights so the solar arrays can be installed in a consistent manner and so walkways can easily connect between them without wasting space; locate rooftop equipment toward the perimeter of the roof so it does not clutter open roof areas.

With a bit of forethought it is possible to locate most rooftop equipment in areas that are unsuitable for solar, thereby maximizing the potential for cost-effectively adding solar. The solar designer can provide recommendations to the architect and engineers that help them identify locations where solar will be most suitable. Have the architect review specifications for the roof system to ensure they support the solar array in terms of durability as well as having light colors to promote lower roof temperatures and corresponding higher solar module output. Including these considerations during the design of the building helps to contain solar project costs by simplifying the eventual design and installation. It will also pay dividends in the form of safe and expedient access to rooftop equipment when maintenance needs to be performed.

The end result of considering the design of the solar project early on in the building's development will be a rational rooftop layout that can accommodate a larger solar array more cost-effectively. A coordinated design will also make it easier to access the roof for inspections and maintenance, and it can make the roof safer by minimizing areas where the roof is congested with equipment. Figure 11.4 illustrates two roof plan layouts: One is optimized for solar while the other is not. You can see that there is less space taken up by walkways in the optimized roof layout, and rooftop equipment is co-located so it takes up a smaller area. The rooftop equipment area also overlaps with the perimeter setback of the solar array. When this off-limits space can do double-duty it leaves more area for solar modules. As this illustration demonstrates, coordinating the roof layout can lead to more efficient and productive solar array designs.

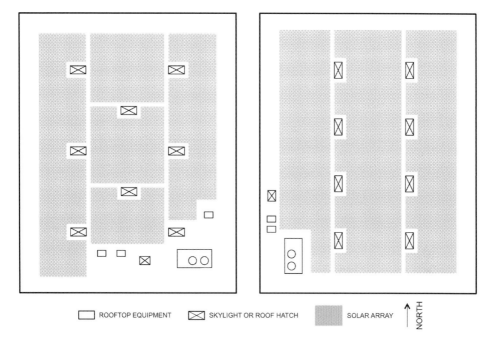

Figure 11.4 Roof layout comparison.

Note: The inefficient roof layout (left) is due to the haphazard placement of rooftop equipment and skylights across the roof. The efficient layout (right) illustrates the same equipment and skylights organized on the roof in locations that reduce the impact on solar module placement. The result is a more efficient solar layout that can accommodate greater solar coverage.

Water management

Water and solar electricity do not mix. Ensure that water drains from the roof quickly and completely. A good solar array design ensures that any water on the roof drains completely through the solar array on its way to the roof drains. Keeping water from accumulating on the roof has several benefits, including:

- prolonging roof life by minimizing standing water;
- allowing solar modules and electrical equipment to dry out;
- minimizing weight on the roof.

Ensure that the solar design does not have large racking anchors or ballast trays that could block the flow of water on the roof. Verify that any large equipment is elevated off the roof or has a curb designed so water will not accumulate behind it. If you see debris from nearby trees accumulated on the roof, consider trimming or removing the foliage to prevent this material from clogging drains. Removing the source of debris will also help keep the solar panels clean so their performance does not suffer. These measures will not only help the roof last longer; they will reduce the possibility of standing water that can damage solar components.

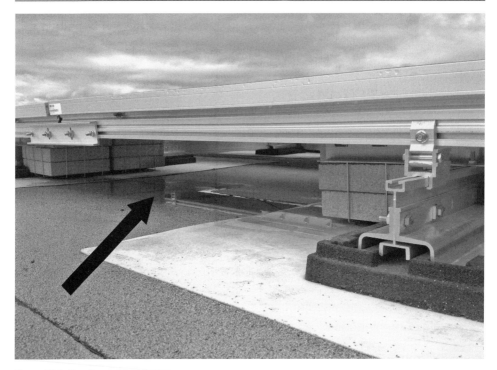

Figure 11.5 Ponding under a solar array.

> **Quick tip**
>
> One inch of water weighs five pounds per square foot; one inch of water covering a 20-foot by 20-foot area weighs more than one ton.

A few inches of water accumulated on a roof is quite heavy. Combining the weight of standing water to the weight of the solar array can over-stress the roof structure. This could create low spots in the roof that lead to even more standing water, accelerated roof deterioration, and damage to solar equipment. In extreme cases it could contribute to the structural weakening or failure of the roof when combined with a severe weather event or seismic activity.

Temperature-related expansion and contraction

Temperature changes that occur from day to day and season to season cause solar equipment components to expand and contract. As this happens repeatedly there is potential risk of stress on the solar array as well as wear in areas where the racking or cable trays contact the roof. A darkly colored roof in a warm climate may start out at a temperature of 50° before the sun rises, and by mid-afternoon its surface temperate could increase to 150° or more, before cooling back down at night. Over time,

Figure 11.6 Roof expansion joint and conduit expansion joint.

Note: In the foreground of this photo the black conduit has a gray flexible expansion joint where the conduit crosses an expansion joint in the roof of this structure. The expansion joint helps to ensure that over time the conductors and conduits do not become over-stressed from expansion and contraction movement in the underlying building.

Figure 11.7 Expansion joint in a cable tray.

Note: The joint of this cable tray allows for expansion and contraction. Note the elongated bolt hole and crimped grounding wire that accommodates the movement within the cable tray.

Figure 11.8 Support blocks and slipsheets under solar equipment.

Note: This conduit and combiner box are mounted on recycled rubber support blocks and placed on slip-sheets to accommodate expansion and contraction, and to minimize roof wear.

expansion and contraction causes wear that eventually breaks down the roof and solar equipment.

Expansion joints and flexible anchors can limit these stresses in conduits, racking, and cable trays. Slipsheets and flexible rubber support blocks can reduce the friction between solar equipment and the roof surface. Light-colored roofs also play an important role in reducing stress and wear. Since white roofs experience much lower surface temperature swings than dark colored roofs, expansion and contraction is reduced significantly. Minimizing the amount of expansion and contraction is a first step to reducing a major factor that contributes to the risk of roof wear and potential system damage.

Structural loads and engineering

Structural design is required to ensure that the racking system is adequate to resist wind and the weight of the solar modules. Structural design is also required to ensure that the building structure supporting the solar array is strong enough. Structural design calculations can affect the choice of solar technology and racking systems if they indicate that the building structure cannot support the solar array as designed.

These decisions impact solar project costs and can affect system output. Just as importantly, limits on structural roof loads can impact the maximum size of the solar

Figure 11.9 Close-up view of a support block and slipsheet.

Note: The slipsheet has been marked to indicate the location of the support on a given date. Over time this will be monitored to measure if the array experiences any movement away from its original position.

array. The impact of this is a smaller array and lower power production at a potentially higher unit cost than would be the case if there were no structural limitations. Proactively identifying and addressing structural limitations will allow you to manage a key risk factor that can increase project cost and potentially reduce the achievable size of the project.

Allowable roof loads

A typical solar array weighs anywhere from three to eight pounds per square foot. The exact weight will vary for each project depending on building code requirements, system design, and equipment selection. Once an engineered design of the solar array is available, have a structural engineer analyze the impact that the added weight of the solar facility will have on your building. If the weight of the planned solar project cannot be safely accommodated, identify ways to modify the project's design to reduce weight or distribute it more evenly. This may mean reducing the size of the solar facility, changing to a different racking system, or using a different solar module technology. The alternative – adding structural reinforcement to the roof – is seldom cost-effective. These changes are likely to affect project economics and should be considered carefully in context of the other goals such as energy output, cost and constructability.

A factor that comes into play when determining how much weight a roof can support is the building structural codes. In most markets these codes have become more stringent over time. Buildings that were originally built under an older building code may have less spare structural capacity to support the added weight of a solar facility. This is because many codes stipulate that if changes are made that increase the building's structural loads above a certain amount, the whole building must be brought into compliance with the current code.

Wind forces

Wind forces, also referred to as wind loads, are one component of the structural analysis that can have a significant impact on solar project design. Wind loads create forces that pull up or push down on the roof. The same forces must also be resisted by the solar array in order for it to remain on the roof. Strong winds can blow solar modules off the roof if they are not properly installed to resist wind uplift. Uplift occurs when wind passing over the edge of a roof creates a zone of low pressure. It is particularly relevant for large, flat rooftops and very tall buildings where wind can act strongly in certain spots. There are several basic ways a solar design engineer can address wind forces. They can:

- avoid zones where uplift is greatest near roof edges and building corners;
- install airfoils and other devices to reduce wind forces;
- install modules at a lower angle to reduce wind forces;
- increase the amount of ballast to hold the array down;
- physically anchor the array to the roof.

Setbacks may already be incorporated into the project to allow for access to parapets and rooftop heating and cooling equipment. If setbacks need to be increased, they can reduce the area available for the solar array. This can result in a smaller and less economically viable project. On the other hand, increasing setbacks may reduce the need for additional ballast – an added expense that also increases roof structural loads. Alternatively, a racking system that relies on physical anchors connected to the roof structure may be required to overcome wind forces. This type of solution may increase project costs. Because the anchors penetrate the roof, there is also an increased potential

Figure 11.10 Airfoils attached to a row of solar modules.

Note: Aluminum airfoils have been installed behind the last row of solar modules to reduce wind uplift for these 30-degree tilt modules. Airfoils were not required for other rows of modules within the field of solar modules.

for leaks if the anchors are not properly flashed into the existing roof. A system anchored to the building is very resistant to wind forces and tends to weigh less than ballasted designs, so it may be able to be installed on more areas of the roof.

The solar project may also include accessories that reduce the force of the wind. Airfoils, curved aluminum cowlings fastened to the racking system, reduce the effects of wind uplift. These are fairly common on large rooftop projects, but adding airfoils to a project can increase costs.

The angle, or tilt, at which the solar modules are installed can also affect wind forces. Steeper tilt angles increase the wind's effect. On weight-constrained roofs, it may be more cost-effective to install modules at an angle that is almost parallel to the roof rather than to tilt the modules toward the sun. The lower angle requires less ballast and can reduce installation costs. The trade-off is that solar module output will be marginally reduced because the sun's energy hits the modules less directly. To minimize wind uplift forces on a solar array, solar modules are often angled from zero to ten degrees from horizontal. Solar modules can be installed at higher angles, up to 30 degrees or so, where wind forces are not a major concern or where the attachment of the solar array can be designed to accommodate higher wind forces.

You will often see a combination of solutions used to address wind uplift on solar projects. Modules may be installed at a low angle and have airfoils as well as ballast to

Figure 11.11 Low-profile airfoils attached to a row of solar modules.

Note: Due to the low 5° angle of the solar modules, the size of the airfoil on the back of the last row of modules is fairly small. This array is affixed to the supporting structure below without ballast.

resist wind forces. Another project may use a combination of ballast and roof anchors to reduce array weight and address wind forces. In other instances, the project may be fully affixed to the underlying building structure, thereby eliminating the need for ballast or airfoils altogether. The impact of wind forces can be addressed effectively for most projects when the structural engineer and the solar design engineer work together to find a solution.

Final thoughts

Careful planning that identifies potential risk factors for the project's design, schedule, permitting, and engineering goes a long way toward mitigating areas that could increase project costs or reduce revenues. The outcome will be a project that meets your expectations and provides years of trouble-free operation. Chapter 12 focuses on aspects of solar project development related to construction planning that support the successful execution of solar projects.

Chapter 12

Construction planning

Chapter summary

- Early in the design consider the way the project site can impact staging the installation work.
- Provide temporary roof protection in areas of high traffic during construction.
- Review staging plans with the contractor to minimize disruptions to existing building tenants.

The construction of a solar facility can impact building operations and affect tenants. The construction process needs to be planned carefully to minimize these impacts. The logistics of delivering materials to the site can be significant – a large project may require tens of thousands of solar modules, electrical equipment as big as a car, and many thousands of additional components. Staging the construction also requires space on the ground that needs to be coordinated with the business activities of building tenants.

The solar facility connects to the building electrical system or the utility grid. The former can affect the building's operations during the interconnection activity, and the latter may require trenching across the site to reach a utility connection point. Effectively managing these impacts during construction is a key aspect of being able to deliver a cost-competitive project that integrates with the building and existing tenants while ensuring trouble-free operation.

Site access

Site access requirements during construction tend to be overlooked until work is about to begin. Waiting until construction is underway can lead to situations where the tenant's use of the site conflicts with the contractor's staging plans. For example, a suburban office building on a large site may support straightforward site access, while an urban high-rise office building is likely to have significant site access restrictions. In some cases these access limitations may dictate solar facility design, such as using off-site prefabrication to speed-up deliveries and installation. If these needs are not identified until late in the project's design, changes can lead to higher costs or delays that affect the construction process.

Communicate with tenants so they understand how they could be affected by the work and are aware of the measures being taken to minimize disruptions. This helps to manage their expectations and provides an opportunity to identify any constraints they may have, such as a peak period of activity that will compete for space on the site, or an important meeting when noise from the construction would be especially problematic. Factor this into the contractor's work schedule.

Quick tip

Roof access through existing roof ladders or stairways is rarely feasible during construction for reasons of both safety and speed. During construction on lower-height buildings, temporary scaffolding staircases are often installed to provide roof access.

Construction staging areas

Some properties have little space available for construction staging areas to support solar project installation. Constrained staging areas can cause tenant disruptions, slow the pace of construction, and in some cases can lead to increased project costs. This risk can be remedied by consulting with the solar developer and general contractor to identify suitable staging areas. Site constraints should be identified and accounted for in the design to minimize any increased costs or delays.

As part of the staging plan, consider areas on the building where pallets containing solar project components will be stored temporarily until they can be installed. If this is on the roof, have the structural engineer determine how many pallets can safely be loaded at one time. Pallets may weigh 1,000 lb or more, and placing them on the roof may require cribbing or other means to distribute the load safely.

In some cases staging or access limitations could make one design solution preferable to another. For example, a solar system designer at an urban site with limited staging areas may choose to have sections of the array fabricated off-site. This would allow large sections of the project to be lifted into place without the need to store unassembled components on-site. Identifying special staging requirements early in the design process will reduce costs and help to eliminate the risk of delays during construction.

Building a rooftop solar array requires space on the site for temporarily storing and staging construction materials before they are loaded onto the building and installed. The contractor will need space to unload delivery trucks and to prepare materials for installation on the roof. For a prototypical project of 100 kW, there could be several large trucks making deliveries to the site each day. This size of project could require as much as 10,000 square feet (930 square meters) of space for this work. In addition to needing space for these materials, the construction project may also require space for:

- cranes;
- lifts;
- job trailers;
- security station;
- workers' personal vehicles;
- contractor's work trucks.

Figure 12.1 Pallets containing solar modules on a roof.

Note: Pallets are spaced out along the length of the roof to reduce concentrated roof loads. Also note the walkway protection mat to the left of the pallets and temporary skylight fall-protection at skylights to the right.

Lessons from the field

Step on the scale: weight limits on the roof

I worked on a solar project in southern California that was being installed on the roof of a large warehouse. The structural engineer identified a maximum safe concentrated load that was allowed on the roof. Because of this, the contractor implemented several job-site requirements. These included:

- placing pallets of materials on long wood cribbing to distribute the weight across several roof structural members;
- mandating an eight-foot gap between pallets to distribute their weight;
- requiring all personnel and visitors on the roof to remain arms-width apart to avoid overloading the roof in a given area;
- enforcing a weight limit of 250 lb (113 kg) for anyone that worked atop the roof.

These rules were diligently enforced – when one of the project's sponsors visited the site and did not meet the weight limit, they were restricted to a rooftop viewing area where the structural engineer had approved higher loads.

Communicate with the solar developer and contractor about your building's operational needs related to the delivery, storage, and staging of construction materials. This helps to ensure that activity is limited to suitable areas of the site. Being proactive about balancing the needs of tenants with the needs of the contractor will enable a more cost-effective construction process and go a long way toward minimizing potential tenant disruptions that could result.

Roof protection

The burst of activity necessary to construct a solar project on a commercial property can, if not managed properly, have a detrimental effect on the roof of the host building. The discussion in this section is focused on activities that take place during the construction process. In a typical year, a roof may see foot traffic from two to four people for a few hours a few times each year, primarily for inspections and minor repairs. In contrast, construction of a solar array could put dozens of workers, wheeled carts, and pallets of materials on the roof daily for several months. In other words, in those few months the solar project can deliver to the roof the equivalent of ten or more years' worth of foot traffic.

The risk of roof damage can be managed effectively by including provisions in the construction contract that require the contractor to provide roof protection. These measures include:

*   requiring that the contractor repair any damage caused by the construction activity;
*   ensuring that the roofing manufacturer acknowledges the solar project and will honor their warranty following the solar project installation;

Figure 12.2 Roof protection (Source: Woody Welch Photography).

Note: There are plywood sheets to protect the roof in the main high-traffic aisle, running from bottom left to upper right in the photo.

Lessons from the field

Gravel-ballasted roofs

A commercial building may have a roof surface covered with gravel or other stone used as a protective surface for the underlying roof system. Removing gravel or stone ballast from a roof prior to the installation of a solar facility brings certain risks that are not present with non-ballasted roofs. Vacuuming up the gravel ballast can cause wear on the roof from worker foot traffic and equipment. Ballast fragments that are inadvertently left on the roof can contribute to roof damage if they are stepped on and damage the roof membrane. This could also occur if any gravel is trapped under the contact points between the solar facility and the roof.

One common solution to circumvent the potential concerns with this type of roof is to replace a ballasted roof with a non-ballasted membrane roof at the time of solar project construction. This has several advantages. First, it provides a brand new roof underneath the solar facility. Second, by eliminating the ballasted roof system the weight of the roof is reduced. This increases the amount of additional weight of solar equipment the roof can support. Eliminating ballast also makes future maintenance easier in the areas where the solar equipment is located since it will not need to be removed in order to perform repairs.

- performing pre-construction and post-construction roof inspections to identify any areas where roof damage or increased wear has occurred;
- requiring the contractor to have a roof protection plan that specifies the protective measures that will be used for the project. This plan may include temporary walk surfaces such as rubber mats, plywood, synthetic turf, or other material placed wherever there will be concentrated foot and wheeled cart traffic. This includes major aisles where workers travel up and down the roof. Paths of travel should also be restricted to these protected areas, as well as areas where workers take breaks. Areas where materials are loaded onto the roof also need temporary protection to prevent damage from pallets of materials.

Solar array layout also plays a part in minimizing the risk of roof wear both during and after construction. Review the solar array design to ensure the layout allows multiple paths to access rooftop equipment. If foot traffic must be concentrated in certain areas, verify that there are walk pads to protect the roof.

Final thoughts

Working closely with the contractor to proactively identify possible construction activities that could impact building tenants or the building will avoid many of the potential risk factors that arise when commencing work on-site. Ensure that there is adequate space on the site for materials and contractors' vehicles. Have the contractor provide temporary protection for high-traffic areas on the roof. These measures are simple and they can go a long way toward making your solar project run smoothly. Chapter 13 introduces considerations that emerge when planning and constructing electrical equipment for your solar array and identifies ways to manage the potential risks that they bring.

Chapter 13

Electrical equipment

Chapter summary

- Electrical equipment is typically housed in an indoor electrical room or installed in a custom-built outdoor enclosure; small systems may be wall-mounted.
- An electrical enclosure provides security and protection for electrical equipment.
- Consider the sound produced by the electrical components of a large solar array when siting the equipment.
- Ensure that the electrical enclosure receives adequate ventilation to prevent equipment overheating.

A solar array is more than just solar modules. A complete project will include equipment such as conduits and cable trays, inverters, switchgear, system-monitoring equipment, electrical transformers, as well as a weather-monitoring station. Space is required for this equipment. In a fully occupied building on a fully developed site the space for this equipment may not always be readily available. Proactively identifying potential sites for this equipment early in the project's design allows any incremental costs or other design changes needed to accommodate the equipment to be factored in. Solar system designers will typically place this equipment on the site or in the building where it is most convenient and cost-effective for their needs. This is usually based on the shortest possible cable runs and proximity to the main electrical feed or utility grid connection. This may not be the best location for the property owner or building tenants. Resolving this potential difference early in the design process will eliminate late-stage design changes that can add risk to projects by delaying the schedule and potentially increasing costs.

Solar projects require an area for electrical equipment either inside an electrical room, on the roof or on the ground near the building. Where feasible this should be near both the solar array and the building's main electric service, but it is not essential. Interior locations provide the best protection and security for the equipment. This equipment generates heat that may need to be exhausted from the electrical room. Equipment installed indoors may require the addition of fire-rated partitions or other fire separations to comply with building codes.

Rooftop and exterior locations will not occupy rentable area or reduce the space available for other building systems. Equipment located on the ground is usually easier

Figure 13.1 Outdoor electrical enclosure.

Note: The vertical painted steel fence at this location allows ample airflow to reach the electrical equipment inside while providing security against unauthorized entry. Also note the enclosed vertical conduit enclosure on the wall that terminates in the electrical enclosure. The chase was designed to blend in with the existing building façade, with matching reveals and paint job.

> ## Quick tip
>
> A solar facility typically requires one to two square feet of space per kilowatt (0.09–0.18 square meters) of system size to accommodate electrical equipment. For a 500 kW solar project this means between 500 and 1,000 square feet (46 to 93 square meters).

to install, maintain, and replace compared to equipment on the roof or inside the building. Identifying adequate space for electrical equipment may become an unexpected challenge on space-constrained sites. As a result this should be addressed early in project planning before it can complicate construction or disrupt tenants.

Electrical enclosure

The design of an enclosure for outdoor electrical equipment typically includes protective walls, screens, and fences that blend aesthetically with the site and building. This space may or may not have a roof, depending on the climate where it is located and depending on what security concerns exist. A well-designed enclosure helps to minimize concerns from landlords, tenants, and adjacent property owners by providing the following benefits:

- protection from tampering, theft and vandalism;
- prevents unwanted entry by people and animals;
- reduces noise outside the enclosure;
- protects electrical equipment from weather.

Quick tip

An electrical enclosure is a dedicated space, usually on the ground or inside a building electrical room, where the electricity-handling equipment for a solar facility is placed. This is where the inverter, transformer, disconnects, and metering equipment are typically located.

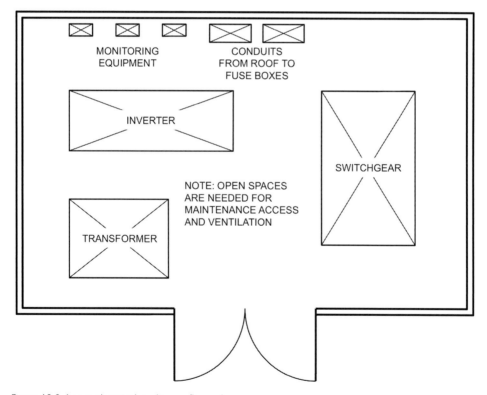

Figure 13.2 Large electrical enclosure floor plan.

Note: This example layout is for a large solar project with a freestanding inverter, switchgear, and transformer. Space has been provided on all sides of the self-contained equipment to facilitate maintenance and repair. Double doors provide easier access with parts and tools for maintenance needs.

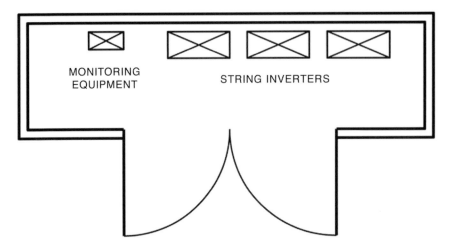

Figure 13.3 Small electrical enclosure floor plan.

Note: This example layout is for a small solar array with several string inverters. In this case, the array does not need a freestanding transformer or switchgear because the output voltage from the inverters is compatible with the voltage of the building it is servicing.

Figure 13.4 Area inside a large electrical enclosure.

Note: The electrical equipment in this space requires a clear area on all sides for safety and ease of servicing. The switchgear is pictured in the foreground, and the transformer can be seen in the background on the right. Also note the security screen overhead to prevent unauthorized access.

Figure 13.5 Outdoor electrical enclosure.

Note: The electrical equipment for this large solar array is protected by a chain-link fence enclosure. This installation does not provide sound attenuation or weather protection.

Sound from electrical equipment

Electrical equipment such as the inverters and transformers generate sound that can exceed 65 decibels. This is a moderate sound level, but it could seem loud in an enclosed space or on an otherwise quiet site. A solid-walled enclosure will reduce sound levels outside the area containing the electrical equipment. If the room is inside the building, masonry or concrete walls are ideal for the enclosure, but sound-attenuating insulation in the walls and sound-insulated doors can also be effective at helping to keep sound from escaping. Increasing the wall height for exterior enclosures will also help to reduce nearby sound levels. Keep in mind that sound will escape from openings in the walls such as gates and doors, so consider solid barriers to reduce sound transmission, as long as necessary airflow for cooling is maintained. These measures will help to ensure that your project doesn't interfere with adjacent property owners or affect a tenant's quiet enjoyment of their space.

Lessons from the field

Advancements in inverter technology

The solar project designer has a range of technologies to choose from when selecting an inverter. A traditional central electrical inverter can be replaced with numerous smaller devices that perform similar functions if there is limited space for large electrical equipment. There are two types of devices: string inverters and micro-inverters. While generally more expensive, these technologies can be installed in smaller spaces and provide a level of redundancy compared to a single central inverter. Maintenance can also be less intrusive since smaller sections of a large project can be shut down without taking the entire project offline.

String inverters handle the electricity from a small portion of a larger array. For example, a 200 kW array using string inverters may use 20–30 such devices. The same system with a central inverter would have a single inverter. Micro-inverters are another alternative. A small wallet-sized inverter is installed on each solar module and acts as a mini-inverter on the panel. These technologies tend to be slightly less efficient and they are usually more expensive than central inverters, but they can nonetheless offer a solution for space-constrained sites that cannot provide the area required for a central inverter.

Figure 13.6 Wall-mounted string inverters.

Note: Six string inverters are shown mounted to a metal frame affixed to a rooftop parapet wall. These inverters were installed on a parapet wall away from direct sunlight to improve inverter performance.

Final thoughts

Working with the solar design engineer to specify the right inverter for your project and designing a space for it that provides protection and safe operation helps to ensure that your project operates reliably and trouble-free. In Chapter 14 another important aspect of many solar projects is discussed – how the roof of your building impacts the design and function of a solar project.

Roofing

Chapter summary

- Ensure there is adequate space around rooftop equipment to allow for inspection, maintenance, and replacement.
- A solar array typically covers 50–80 percent of the roof area. Its coverage is limited due to setback requirements and zones where modules are not suitable.
- Consult with a structural engineer to ensure the roof system is capable of supporting the weight of the solar array.
- Avoid installing solar on a roof that will need to be replaced during the lifetime of the array.
- Design in flexibility and leave space in areas where tenants may want to install rooftop equipment in the future.

The roof of a commercial building is often a preferred place to install a solar array. It is essential to consider how the roof and solar array interface in order to ensure a long and trouble-free life for both. Premature failure of the roof could lead to costly repairs as well as the temporary dismantling of the solar array to provide access to the roof. Solar facilities can be installed atop most roofing systems as long as the solar equipment is designed appropriately and the roof is in good condition. Regardless of the building's age, a roof that is new or nearly new will provide the best base atop which to install a solar project. The roof should be expected to last as long as the solar project – typically 20 years or more.

When a new roof is being installed it provides the perfect opportunity to implement a few modest changes to prepare the roof for solar. Common improvements that make roofs more suitable for solar projects include:

- eliminating low spots where water ponds;
- increasing roof slope to promote positive drainage;
- installing additional scuppers or roof drains to provide enhanced roof drainage;
- replacing skylight and equipment curbs to eliminate a common source of leaks;
- upgrading skylights with fall protection;
- installing a recover board to provide a rigid roof underlayment.

Skylights and mechanical equipment

Skylights and equipment on the roof will need periodic inspections and maintenance during the life of the solar project. Replacing a roof curb under mechanical equipment can be difficult on a wide-open roof, never mind a roof that is blanketed with solar modules. Proactively replacing or upgrading curbs and flashings before the solar array is installed can limit repairs during the life of the solar array that could be more costly if the work needs to be done within a field of solar panels.

Recover board

Many commercial buildings have roof membranes that are installed atop rigid insulation board. Depending on its density, this foam board insulation may not be well suited to supporting concentrated loads for long periods of time – as would be the case for a solar array sitting on the roof. The insulation can become compressed over time, leading to low spots. These depressions may also stretch the roof membrane and contribute to premature roof system failure. Installing a recover board, a thin, rigid panel on top of the insulation, is one solution that enables the roof surface to distribute the loads imposed by the solar array without causing excessive wear on the roof membrane or insulation. Including this type of recover board in the roofing specification may add some cost, but it will enhance the ability of the roof to handle the weight of a solar array. Higher-density insulation may also be considered.

Figure 14.1 Setback at a skylight.

Note: This photo illustrates the separation between the solar modules and a skylight. This open area provides space for maintenance work on the skylight without affecting the solar array. Also note the aisle extending to the right between solar modules to allow access to nearby skylights.

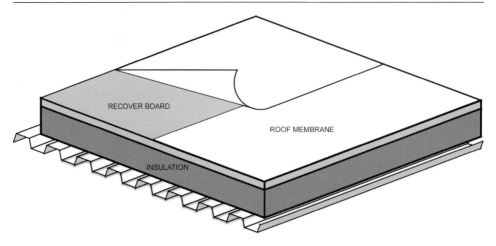

Figure 14.2 Cross-section of a roof showing recover board.

Lessons from the field

Does roof color matter?

Darkly colored roofs heat up more than lighter colored roofs. Higher temperatures reduce solar module electricity output. Darkly colored roofs can be up to 70°F (21°C) hotter than white roofs on sunny summer days. This temperature difference can reduce module output by up to 10 percent, because module performance declines as temperatures rise.

In cold climates that experience regular snowfall there are arguments to stick with dark-colored roofing materials. Sunlight heats up dark roof surfaces and speeds up snow melt. This quickly melts away any snow built up around the solar modules, allowing them to return to full production sooner. An added bonus in cold climates is that solar modules produce incrementally more energy because of the lower temperatures. Keep in mind that solar modules installed in warm, sunny regions such as southern Europe or the southwestern United States will produce significantly more electricity than in colder, more northerly climates, regardless of roof color. Even so, having the right roof color will improve system output regardless of location.

Addressing existing roofs

Constructing a solar project on a new roof is almost always preferred, but it may not always be feasible. Installing an array on an older roof introduces several considerations that can affect the underlying building and the profitability of the solar project. If the roof will require replacement at some point during the lifetime of the solar array, removing and reinstalling the solar equipment on the roof will add significantly to the cost and complexity of the project. On a large building roof with a large solar facility this could mean several months of downtime. There will be a direct cost in labor and time, plus costs related to lost electricity revenue while the array is offline. If the system has been in place for a long time, replacement of wiring or other equipment degraded by outdoor exposure could push costs even higher.

If the solar project must be taken offline it is usually better to do this in the winter months when both solar energy output and energy prices tend to be lower. This helps to minimize loss of energy sales revenue. It may also be preferable to do this work in phases to minimize downtime and to spread out the cost over more than one year. Because of the cost, complexity, and disruption, replacing a roof should be considered a last resort during the operational life of your solar project.

Where starting with a new roof is not feasible, include provisions in power purchase agreement (PPA) or roof lease contracts to address the expected re-roof scenario. Where the cost of removing and reinstalling the solar facility is borne by the solar facility owner and not the property owner, the cost should be factored into their financial model for the project. This will translate into less revenue that can be shared with you in the form of PPA energy pricing or lease rates. That is another reason why re-roofing is often best addressed proactively so it does not put an unnecessary drag on solar project value.

Quick tip

Solar modules installed on a roof will block much of the sun's ultraviolet light; a key contributor to roof degradation. Over the life of a solar facility you are likely to see reduced roof UV damage in areas covered by the solar array.

Ballasted racking systems

Attaching the solar array to the building is a key point of ongoing operational risk. A properly designed and installed array will provide years of trouble-free operation while a poorly completed project could create new maintenance headaches and make existing ones worse. Solar arrays are either mechanically fastened with bolts or screws to the building structure, or ballasted to the roof. Ballasted systems are most common on low-slope roofs and in locations that are not subject to extremely strong winds. The racking system separates the solar modules from the roof and holds them in the correct position to optimize energy production. The connection of the racking system to the roof enables the solar array to resist forces that could damage solar equipment, such as:

- wind uplift;
- seismic shaking;
- movement due to vibration, wind, or other forces;
- expansion and contraction due to temperature changes.

Ballasted racking sits atop the roof, held in place by metal or plastic pans that are typically filled with concrete blocks or stone. An engineer will specify the amount of ballast at each location to ensure that the correct weight is used based on the design

Figure 14.3 Concrete ballast blocks installed on a roof.

Note: Ballast blocks directly support the racking system in this design. The ballast blocks sit atop rubber slip-sheets.

Figure 14.4 Ballast basket mounted to a racking system.

Note: Rubberized pads and white PVC slipsheets have been installed under the racking system to protect the roof. In this design the ballast blocks are held in metal baskets mounted to the frame of the racking system.

Lessons from the field

Ballast weight

The weight and location of ballast is carefully engineered. Ballast holds the racking system onto the roof to resist wind forces that would otherwise try to pick up the solar modules and blow them away. Wind has more of an effect on some parts of the solar array than others, which affects the amount of ballast needed in a given location. More ballast is usually specified at the corners and edges of the solar array than in the center. While the roof structure may support the "average" weight of the entire solar array, additional ballast on a particular area of the roof must be carefully examined by the structural engineer to ensure it does not overload structural members in that location. If it does, this could necessitate changes to the design, such as:

* mechanically attaching racking to reduce ballast weight;
* relocating the solar array to roof areas that will support greater loads;
* spreading out the array to reduce concentrated loads;
* reinforcing the roof structure.

For example, assume that the average weight of a ballasted solar array is six pounds per square foot. Where uplift forces are greatest the engineer may specify array weight of eight pounds per square foot. In the middle of the array where uplift forces are weakest, this may be reduced to as little as three or four pounds per square foot. Roof structural members that can support up to five pounds per square foot may not be able to support eight pounds per square foot. If this is the case, changes to the design may be required.

of the solar array. The engineer must also ensure that the roof structure can support the weight of the solar array plus the ballast. Ballasted systems rely primarily on the friction between the roof and the racking to keep the solar array in place. Even systems that are fully ballasted will often have a limited number of mechanically attached connections to ensure that the system stays where it is intended.

Mechanically attached racking systems

Racking systems can be mechanically attached to the roof structure with screws, bolts, or adhesives. A common solution is to anchor short, round pipe posts to the roof and then attach the racking to them. Using mechanical anchors typically results in a more lightweight solar array because there is little or no need for ballast. Anchors may also be used in conjunction with ballasted projects to hold the array in place on the roof to resist slow movement or sliding due to wind, vibration, or expansion and contraction. Experienced roofing contractors can install roof penetrations for this purpose based on

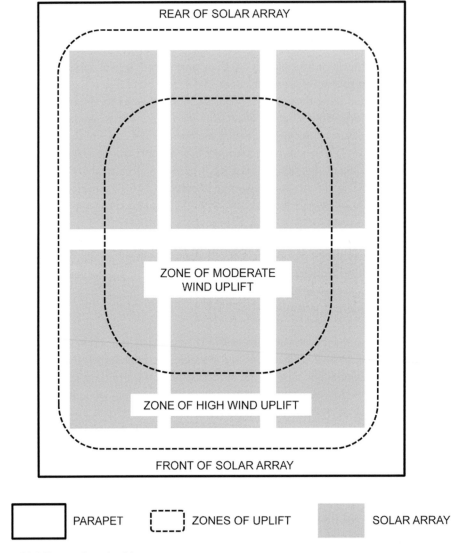

REAR OF SOLAR ARRAY

ZONE OF MODERATE
WIND UPLIFT

ZONE OF HIGH WIND UPLIFT

FRONT OF SOLAR ARRAY

PARAPET   ZONES OF UPLIFT   SOLAR ARRAY

Figure 14.5 Zones of wind uplift.

specifications provided by the project engineer and the roof manufacturer's approved details.

Roof penetrations create places where there could be a leak in the future. A large rooftop solar project could require hundreds or even thousands of these anchors. In these cases it is important to:

- use an experienced roofer for the flashing work – do not rely on the solar installer;
- install penetrations only on a roof in good condition and with good drainage;

Figure 14.6 Round pipe racking mechanically attached to a roof.

Note: Vertical pipe stanchions have been bolted to the roof structure, covered by the roof membrane and flashed in to provide a watertight seal.

Figure 14.7 Racking system anchored to a roof and sealed with pitch pockets.

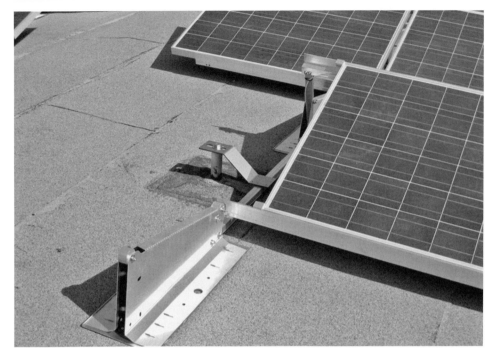

Figure 14.8 Close-up view of a post anchor.

Note: This post anchor, in the center of the photo, functions to prevent the solar array from slowly moving out of position over time. The bent metal clip connects the rail of the racking system to the post anchor.

- specify flashing details approved by the roof manufacturer;
- use standard flashing details to ensure proper installation.

Quick tip

Use flashing details that roofers in your market have extensive experience with. The roofer's familiarity with the detail will help to ensure they are installed properly. Standard details will also make maintenance and repair easier and less expensive in the future.

Roof deflection

Adding the weight of solar equipment on top of an existing roof can cause slight changes in the deflection in the roof's structure. This change may not be immediately noticeable, but can lead to low spots, ponding, wrinkling of the roof membrane, or other changes that did not previously exist. Membrane roofs that have a rigid recover board directly underneath the roof membrane are less susceptible to this in localized areas, since the rigid surface helps to spread out the new loads and reduce compression of the roof.

Have the solar project engineer, roofing manufacturer, and roofing contractor work together to find ways to minimize the impact of roof deflection.

Roof durability

There are several ways to ensure that a roof, whether existing or new, will have the durability and longevity to support solar. These include the following:

- Inspect and maintain the roof regularly.
- Install a more durable roof system.
- Apply roof coatings to protect against wear.

Ensure that the roof has been inspected and proactive maintenance has been done to prepare the roof for the solar installation. Include provisions in PPA or roof lease contracts that specify who is responsible for ongoing roof maintenance, and check with the roof manufacturer to ensure that the solar project will not jeopardize the roof warranty. Fortunately, solar modules are typically placed on the open expanses of the roof. These areas typically require less maintenance than more leak-prone parapets, skylights and equipment curbs. Inspect the roof shortly after rainfall to identify any areas where standing water exists. Standing water reduces roof life and may void the roof warranty. Water creates the potential for damage to the solar equipment and adds weight to a roof that also has to support a solar array.

Where a building's roof can be replaced prior to the solar installation, specify a roof that will have long-term durability to support solar equipment. The roof must provide adequate resistance to wear during installation and during ongoing maintenance activity while resisting the stresses placed on it by the solar array racking system. While it is not required, you may be advised to increase roof specifications by upgrading from a standard type of roof system to a more durable one. This could include using a thicker roof membrane, or specifying a heavier gauge of metal roofing. The main trade-off will be increased cost, but the increased durability can reduce maintenance costs and the potential for leaks well into the future.

Where it is not feasible to install a new roof before the solar installation, consider roof coatings to extend the roof's service life and protect against damage from construction of the solar array. Coatings range from inexpensive paint-like products that provide limited roof protection benefits to reinforced acrylic emulsion systems that act like a new roof membrane and carry long-term warranties. Coating a roof that has a dark surface with a white coating can not only extend roof life, it can reduce rooftop temperatures and improve solar array output.

Roof utilization

Leave space in and around the solar array for the installation and maintenance of equipment used to condition the building and to meet the operational needs of tenants, both today and in the future. While you may end up with a slightly smaller solar array, you will benefit from greater flexibility to accommodate changes in the building's use over time. This will help to avoid relocating portions of the solar array or facing added costs to squeeze equipment into locations that are not ideal.

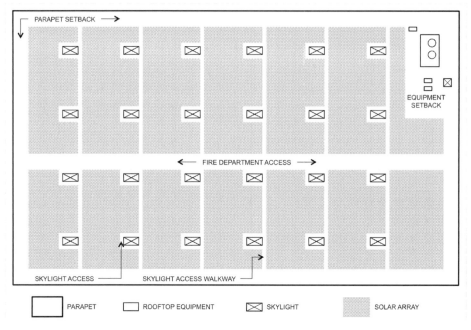

Figure 14.9 Roof plan showing setbacks and equipment access.

Note: This illustration highlights common setbacks for rooftop installations. Note the access on all sides of skylights and rooftop equipment. Fire department access is provided above primary roof structural members for the full length of the building.

Figure 14.10 Setback from an adjacent façade and roof.

Note: The solar array is set back from the adjacent window wall and from the parapet in the rear of the photo. This allows easier access for maintenance and inspection of both the roof and the solar array.

Figure 14.11 Narrow roof edge clearance.

Note: There is relatively narrow roof edge clearance on this side of the solar array. While it is code-compliant, it provides little safe working area for maintenance and inspection of the solar array, roof, or rooftop equipment.

In many cases only a portion of a rooftop will end up being covered by solar modules. This is because open space is needed for walkways, setbacks from the roof edge, and to avoid areas required for maintenance of existing rooftop equipment. Shadows cast by adjacent buildings or trees can restrict where solar panels can be placed. Roof space may also be left open and reserved for future tenant needs. Similar lessons apply to solar arrays and solar electrical equipment located on the ground close to electrical infrastructure, cooling towers, and service areas. Commercial rooftop solar arrays that "fill" a roof will actually often only utilize 50–80 percent of the gross roof area. Many solar projects occupy an even smaller footprint than that. The end result is that the solar facility is going to be smaller than the area of your roof – in some cases much smaller than you may have imagined. Even so, remember to account for the space needed to provide for tenant needs and other equipment that building systems may require in the future. Figure 14.9 illustrates a solar array layout on a roof and indicates several common requirements that limit where solar modules can be placed.

Final thoughts

The roof is the key point of connection for many solar projects on commercial buildings. Careful design of the solar array must take into account the setbacks and open areas to ensure a safe and accessible roof. The roof needs to be protected during construction to avoid accelerated wear or damage. Following these guidelines will help ensure a solar project that treads lightly on the host building. Chapter 15 examines the ongoing operational needs of solar projects and identifies ways to mitigate risks and ensure trouble-free operation.

Chapter 15

Operations

Chapter summary

- Solar systems can operate for up to 30 years when properly maintained.
- Clean solar modules regularly in climates that do not get frequent rainfall.
- Have the solar array inspected periodically and repair any damage.
- Communicate proactively with building tenants about any solar project-related impacts.
- Be aware of hidden costs such as property management administrative time, marketing costs, and even bankruptcy of the solar operator.

Solar projects can operate reliably for as many as 30 years, as long as they are maintained properly. Arrays on buildings typically have no moving parts, and they do not require a source of fuel other than access to the sun. Regular inspections, cleaning and maintenance are nonetheless required to ensure the optimum performance of the array throughout its lifetime. The first year of operation may require regular maintenance as the system is fine-tuned and as individual components complete a cycle of seasonal changes. After this break-in period, operations and maintenance work is generally limited to ensuring that the modules remain clean and that the remaining system components continue to operate as intended.

Many solar project developers provide ongoing operations and maintenance (O&M) services for an annual fee. There are also companies that specialize in providing O&M services for owners of solar arrays. Have these companies provide an O&M proposal that explains the scope of their services, the frequency of their inspections, and what costs are included or excluded from their contract.

In the first year of operation the solar array will be exposed to the outdoor environment for the first time. Changes in weather and temperature may cause some modules to fail if they were installed incorrectly or if they had a manufacturing defect. Any damaged modules should be replaced promptly so they do not reduce overall system performance. Most equipment is warranted to operate for many years, so service may often be available in the event of a problem with an individual component.

Modules need periodic cleaning to maintain their optimum output. Dirt, dust, bird guano, and other substances that adhere to modules are referred to collectively as "soiling." In regions that receive regular rainfall, supplemental cleaning may not be required to remove soiling. In arid regions, cleaning may be required every few months if there is constant dust and dirt but little or no rainfall. More frequent cleaning may be required if the modules are installed parallel with the roof rather than at an angle, since dirt will tend to collect more quickly.

Soiling on panels can reduce output by up to 10 percent if left unaddressed for extended periods of time. The decision of how frequently to clean modules is made to balance the value of lost energy production due to soiling with the cost of cleaning. Keep in mind that the more cleaning that is required means there will be more foot traffic, hoses, and equipment on the roof. That could cause increased roof wear or damage. Where feasible, a connection for water should be supplied near the solar array to simplify the cleaning process and reduce the amount of equipment – such as tanks of water, pumps, and washers – that need to be brought to the roof.

Solar array maintenance

Regular inspections and maintenance will help ensure that the solar array is delivering peak performance throughout its lifetime. Reducing the frequency of inspections or relying too heavily on remote monitoring could lead to:

- accelerated system wear;
- decreased array performance;
- potentially hazardous conditions.

If left unrepaired, these conditions could end up jeopardizing both the solar array and the building host. Fortunately these risks can be managed effectively. Inspection

Figure 15.1 Extensive soiling on a solar module.

Note: The dust covers the entire solar module as well as there being a concentration of dirt at the lower edge.

and service visits are typically conducted about twice a year unless there is a need identified, often through the remote system monitoring, which warrants more frequent inspection or repair. Any loose electrical connections, deteriorated components, broken solar modules, or other repairs will be fixed or scheduled for repair. The solar modules will also be evaluated to determine whether or not they need to be cleaned.

Trees or other tall vegetation near the edge of the solar installation project can create the potential for several negative impacts:

- Trees could cast shadows on some modules when the sun is low in the sky. Over the 20-year life of the project the trees could become much taller.
- Leaves or other debris that falls off the trees could land on the modules and reduce performance. In Figure 15.2 there is a small amount of debris on the concrete apron where the solar modules are located near adjacent trees.
- In the event a tree should fall over or drop a large branch, there is the possibility that it could damage the solar array.

Managing tenants

Tenant rights

Building tenants pay many times more for any given amount of floor area than you are likely to receive from the solar project. For property owners, it is always a priority to

Figure 15.2 Solar array with adjacent trees.

Lessons from the field

The importance of inspections

Solar projects are safe and reliable when installed properly. There are, however, instances where solar projects malfunction. These cases have been primarily due to one of two factors:

1 improper installation;
2 premature deterioration of system components.

These malfunctions lead to poor system performance. In a few cases, these have led to hazardous conditions such as electrical short circuits. Water accumulating in and around electrical equipment is often a contributing factor. These cases are uncommon, but they reinforce the importance of regular inspections to identify conditions that could become unsafe.

Figure 15.3 shows the extreme deterioration of a metal conduit that was run directly on the roof surface. Corrosive conditions including ponding water and bird guano caused the conduit to corrode extensively. Regular inspections would have identified this condition before it became hazardous.

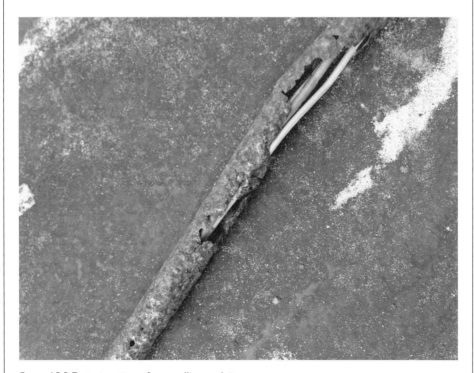

Figure 15.3 Deterioration of a metallic conduit.

ensure that solar efforts do not jeopardize relationships with valuable building occupants. A well-executed solar project can be a source of savings for tenants and revenue for the property. A solar project that is poorly implemented has the potential to negatively impact current and future tenants, just as a disruptive solar construction project or a roof leak that damages tenant equipment can strain your relationships. The first step to avoid this situation is to understand what each tenant's lease includes for rights to use the property. For example, some tenants may have exclusive rights to certain areas of the parking lot and site, while others may share these spaces equally. In other cases the landlord may control the spaces.

Review the contractor's project schedule and staging plan prior to the start of construction to identify possible conflicts with the operational needs of building tenants. Staging areas for solar project construction should respect tenant lease rights and seek approval to use space that is assigned to a particular lease. This also applies to the roof, where some tenants may have rights to use space for their equipment. Tenants may in some cases share roof maintenance costs; adding a solar array could lead to concern that costs borne by the tenants will increase. In these situations it is helpful to have a solar agreement that allocates a portion of the roof maintenance costs to the solar project as a way to demonstrate to tenants that they will not bear additional costs. Figure 15.4 illustrates this arrangement.

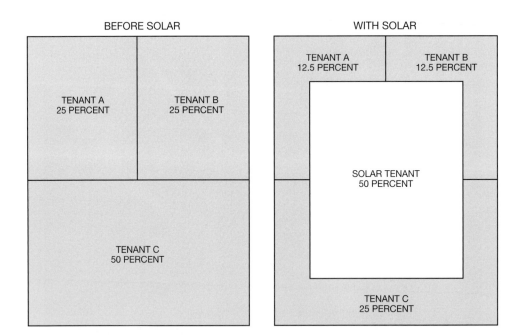

Figure 15.4 Tenant maintenance cost allocation.

Note: This diagram illustrates the allocation of roof maintenance costs in a multi-tenant building before solar (left) and after solar has been added (right). By assigning roof maintenance costs to the solar system operator based on the area they occupy, the proportionate share assigned to the other tenants decreases. This modification can help alleviate concerns tenants may have about higher roof maintenance costs.

Proactive communication

The solar facility construction process could affect tenants by creating noise and dust or limiting parking. Communicate openly with building occupants so they understand the extent and duration of the project. No one likes surprises; notify tenants of any planned power shutdowns and major equipment delivery schedules. This is particularly important if contractor parking, deliveries, or work areas will affect the tenant's use of the property. Trenching across driveways, landscaping restoration, noise from trucks or cranes, as well as visual impacts could also occur during construction. Coordinate in advance if access to tenant space inside the building is required, such as for electrical interconnection work. Inform tenants who they should contact with any questions or problems that come up related to the solar project during both construction and operations. This may be a property manager, a representative from the solar developer, or another person on your staff.

Dust and dirt

With work going on around the building there is a possibility of generating dust and dirt. Particularly if the underside of the roof in an older building is exposed, construction activity could cause dust built up over many years to fall on occupied tenant spaces below. The best solution is a good communication strategy that lets tenants know that this is a possibility and that if it occurs it will be short-lived. Request that the contractor minimize activities that vibrate or shake the roof and dislodge dust. Where this is unavoidable, temporary plastic sheeting can be put in place above tenant workspaces to protect them. Hiring a cleaning crew to come in after each day to clean up any remaining dust may help minimize concerns. Alternatively, you could provide an allowance to tenants to provide their own protection from dust. This provides them with an incentive to provide the amount of protection they believe is sufficient instead of relying on your attempts to intuit their needs.

Security

A solar array can be the target of theft or vandalism. This risk can be managed with careful planning, but it cannot be entirely eliminated. A handful of solar panels can easily cost many thousands of dollars and there could be additional damage from a hasty theft. Scores of modules can fit into the back of a waiting truck late at night. The metal in heavy-gauge wiring, such as copper and aluminum, can be an attractive target as well. Row upon row of shiny solar modules present an enticing target for adolescents with BB guns, baseballs, and rocks. Insurance is one solution to this risk but high deductibles are likely to make many smaller losses fall outside the scope of insurance coverage. Commercial solar facilities are monitored for performance, and missing or damaged modules can be diagnosed, but by then it is too late to prevent theft or vandalism.

The first deterrent to theft and vandalism is to make your system as inconspicuous as possible: out of sight, out of mind. Installing the system so that it is not visible from street level or adjacent buildings will reduce the likelihood that it becomes a target. If portions of the system are visible, find ways to screen them or integrate them into the design of the building so they are less easily identified. For example, planting trees

---

Lessons from the field

Losing control during construction

A hidden gap in property owner control can creep into solar project construction agreements. This can occur when the property owner is not directly financially invested in the project, such as happens when the property owner is utilizing the power purchase agreement (PPA) or lease structure. The solar project developer has a contractual relationship with the property owner, and a separate relationship with the contractor building the project. The property owner, however, has little direct financial control over any party since they are simply providing the space to host the solar array or purchasing electricity once the project is complete. The result can be a situation where the property owner has a concern with some aspect of the construction activity but they lack financial leverage to ensure a solution.

For example, the property owner may see that the project has fallen behind schedule because the solar developer has not gotten materials delivered on time. The property owner has little ability to require expedited delivery or other corrective measures, even if it will affect the project's completion date and encroach on a tenant's busy season at the property. Remedies for this control gap should be addressed with the solar developer to ensure that the construction contract includes appropriate protections for the property owner. Waiting to address this until there is a problem leaves the property owner with little leverage to ensure a satisfactory solution.

---

along the street could be enough to block the view of the solar facility. As long as there is sufficient distance to the trees there will not be any shadows on the modules.

Making it difficult to gain access to the solar array is another effective way to deter theft or vandalism. There are many simple steps you can take to make it more of a challenge to access the solar project, including:

- securely locking fences, gates, and roof ladders. Relocate ladders away from areas that are difficult to monitor effectively;
- trimming trees near the building so they cannot be used to reach the roof;
- avoiding leaving trucks, trailers, or other tall vehicles parked adjacent to the building; they can be climbed on to reach the solar array;
- installing or enhancing site lighting around the building.

In addition to making the solar array difficult to see and reach, the equipment can also be designed to make theft less likely. Tamper-resistant mounting systems deter and slow down criminals. Inter-connecting modules and racking also makes it more difficult to remove panels individually. Install fences, or preferably solid walls, or other physical barriers to protect electrical equipment and wiring from theft and vandalism. Locating electrical equipment inside the building rather than outside is a highly effective deterrent.

Figure 15.5 Gated metal electrical enclosure.

Note the metal grating that protects all sides of the electrical enclosure, including overhead, to prevent unauthorized access. A perforated metal screen has been added to the walls of the enclosure to add further protection. Also note that this electrical enclosure is located on an access-controlled area of the building site, away from major roads.

In cases where theft is a major consideration, the design engineer may specify a solar module that is larger and heavier in order to make them difficult to remove without specialized equipment such as a crane. These specialized products will also have less resale value for thieves. While most solar modules weigh 40–80 lb and can be handled by one or two people, some specialty products can be much larger and heavier, weighing up to several hundred pounds. These modules must be lifted into place by a crane. While modules of this size are rarely used on buildings, they may be an effective solution for carport structures or ground-mounted systems that are in more easily accessed locations on the site. Figure 15.6 illustrates a range of common commercial solar module sizes.

Solar modules are not the only component of a solar array that has theft potential. Electrical components, such as heavy-gauge copper wiring, copper bus bars, aluminum components, and other materials are also at risk. High commodity prices have made the theft of copper and aluminum cables an attractive target for criminals. These cables, more accurately referred to as "conductors," should be run in cable trays and conduits to make them less visible and more difficult to access. Also consider routing conductors underground or in locations that are not readily accessible so they cannot be pulled out easily. Construct a masonry or steel enclosure around electrical equipment at ground level and ensure it is securely locked at all times.

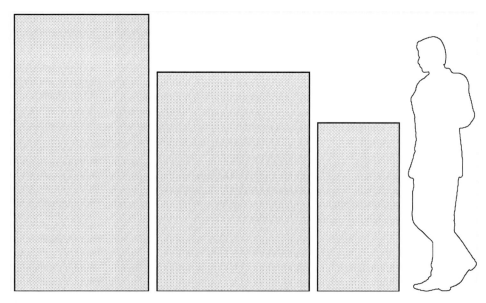

Figure 15.6 Solar module sizes.

Note: Several common commercial solar modules have been compared side-by-side with a 5'10" (1.78 m) person.

Risks of theft and vandalism can be minimized effectively through proper solar project design, security, and monitoring. You can reduce temptation by keeping the system out of sight and by limiting access. These measures will help reduce system downtime and lost revenue, and will help reduce maintenance costs. But this risk cannot be entirely eliminated for solar projects. Over time, the theft of solar modules could become more common as demand for equipment grows. As the industry matures, the relative level of cost and risk that theft and vandalism presents will become clearer for project owners.

Hidden costs

Direct financial commitments under each of the three project structures can be readily evaluated. Less obvious are potential hidden financial obligations that reduce the value of your solar project. Even if the project is being handled in a structure such as the PPA or lease, where there are few out-of-pocket costs for you to bear, unexpected costs may arise outside the scope of the solar project that are necessary to facilitate the installation. These costs erode the value you receive from the project. Anticipating the most likely hidden costs will alert you to the possibility that they could occur so you can take steps to mitigate them.

Property management time

Property management staff time is required to manage aspects of the solar project. This at a minimum includes the time required to administer the lease and periodically coordinate solar maintenance personnel. The investment in time is likely to increase if

---

Lessons from the field

How to steal from a solar project

A commercial-scale solar array will contain thousands of pounds of valuable copper and aluminum. Much of this is in heavy-gauge conductors that connect the solar modules on the roof to the main electrical room. Thieves have been known to show up at a site at night, back a truck up to the main electrical panel next to the building, tie the end of the conductors to the hitch of their truck, and simply drive away. This rips the heavy-gauge conductors right off the building. They then collect the wire and drive off within minutes after showing up on site. This can occur both at completed solar projects as well as at sites where solar array construction is still ongoing and the site is not yet fully secure.

---

the solar project is providing energy for the building. Billing and reconciling the pass-through of solar electricity can take up the time of property management staff. While this may not have a direct out-of-pocket cost, it can increase the burden on existing property management resources.

Default provisions

Some hidden costs may appear only when there are other issues that trigger them. These hidden obligations can lead to complications for the property outside of just the solar project. One example is a requirement in some PPA agreements that the property owner must maintain a minimum credit quality, or they could be in technical default of the PPA contract. This provision safeguards the PPA lender's interests by reducing the risk of default on the power sales contract. Even if you continue to honor the financial obligations of the PPA, a drop in your company's credit rating could trigger technical default provisions in the contract and require you to take corrective action such as posting collateral or paying a penalty. A change in the credit rating of your company may be outside of your direct control, but it could have an unintended and costly impact. Identify these contract provisions and negotiate them to find a middle ground that provides reasonable comfort for you and adequate security for the PPA lender.

A similar scenario could also occur if a property with a PPA contract is sold to another property owner that does not have comparable creditworthiness to the original owner. This could put the sale in jeopardy or trigger technical default on the PPA for the new owner. While hopefully the resolution to this is as simple as talking to the PPA provider and having them accept the assignment to the new owner, you do not want to be forced to address this contract term in the PPA at the moment the sale of the property is hanging in the balance. Ensure that you understand what your PPA contract says about the thresholds for remedying changes in property owner creditworthiness. In addition to your own company's credit, consider what a likely profile of a buyer of the property would look like when negotiating the credit requirements in the PPA. Ensure that the PPA owner cannot unreasonably block or delay the property transaction.

Solar project owner bankruptcy

Under a PPA or roof lease, a bankruptcy of the rooftop solar project owner or operator is a potential source of hidden costs. Bankruptcy could impact the delivery of power from the solar facility, hinder compliance with regulations, and degrade overall system upkeep. If bankruptcy occurs near the end of the PPA or roof lease term, the property owner could be left with an abandoned and devalued solar asset that they ultimately have to pay to remove from their building.

While these risks cannot be eliminated, they can be managed by carefully selecting solar companies that have a long operating history and strong finances. Many solar contracts are sold to investors by the solar project developer; include minimum asset or credit requirements for any future project owners. Roof lease and PPA language should transfer all lease obligations to any new owners. The contract should also spell out what happens in the event of a failure to maintain the solar project's obligations under the lease, with accompanying remedies that could trigger system removal or assess financial penalties. The best way to manage the risk of solar project owner bankruptcy is to have a well thought-out contract that provides adequate protections to shield the property owner from the fallout.

Management and marketing costs

Project management and communication responsibilities by the building owner are often overlooked hidden costs. These include communicating with tenants, lenders, and investment partners, as well as your own colleagues. At a minimum this requires the time and attention of the property owner or their employees. This can reduce the time available for other projects. While this cost may not appear as a separate line item in the owner's project budget, it is a cost in time and manpower that should be acknowledged.

Marketing the solar project can also have a cost that is not initially considered. Marketing is often an essential part of extracting additional value from your solar project. Just like project management time, any marketing effort will have an associated cost for hosting tours, printed materials, or installing informational kiosks in the building lobby. If these personnel and marketing impacts are not identified and managed effectively, they could become an unexpected cost that weighs down an otherwise successful project.

Electricity purchase requirement

The obligation to purchase electricity in a PPA contract may also lead to hidden costs. In many PPAs if the property owner is unable to re-sell energy purchased through the PPA they nonetheless still have to continue purchasing electricity from the solar project. This creates a situation where the property owner may be purchasing more electricity than they can use or sell to tenants. This scenario could happen when it is most damaging to the property owner, such as when the property has recently lost a large tenant or when the property market is in an economic downturn and vacancy is elevated for a prolonged period of time.

The power purchase obligation with no corresponding sales channel can quickly erode the property owner's benefit from a PPA contract. In some cases a single year of lost power sales revenue on the part of the property owner can erase the profit expected from the next several years of solar power sales. The likelihood and potential cost of the power purchase obligation should be considered carefully before any PPA agreement is signed. Aside from mitigating this risk by having multiple tenants purchasing electricity or using it for common area needs, you could also negotiate contract language that suspends or reduces the purchase requirement if there is an extended vacancy. If this risk is a major concern, consider reducing the size of the solar array to shrink your exposure to lost energy sales revenue.

Roof maintenance

Roof upkeep and other operations and maintenance costs can be affected by having solar on the roof. This can be the case even if a third party in a PPA or roof lease has responsibility for a portion of the roof's upkeep. These costs may include more labor-intensive roof maintenance procedures. Consider the roof contractor that has to walk around a solar array with his gear in tow in order to reach each area of the roof that requires maintenance. The increased time required could result in higher costs even if the repair does not relate to the solar array. Ensure that the design of the solar project includes walk paths around the solar array with direct access to rooftop equipment, parapets, and skylights.

Final thoughts

Effectively managing physical and operational risks during planning and construction is an important and often overlooked step toward ensuring a successful outcome for your solar project. Proactively addressing these risk factors will benefit your project by reducing the likelihood of delays, increased costs, and disruptions to building tenants. The risks identified in this chapter can be anticipated as the project takes shape, and steps can be taken to mitigate their impact. Your control over the operations of the solar project will vary depending on the type of solar project structure pursued. As a result, be sure to verify potential tenant impacts, as well as internal time and resource commitments. Chapter 16 continues the discussion of risks to the profitability and value of your solar project, but it introduces a second category, systemic risk factors, which lie outside the direct control of the project developer or the property owner.

Systemic risks

Chapter summary

- Solar projects face systemic, or industry-wide, and regulatory risks that can affect the future value of solar projects. These risks cannot be entirely identified or controlled.
- Incentive programs are subject to changes that can affect the attractiveness of solar projects.
- The level of supply and demand for solar host sites in a market can affect the value commercial properties receive.
- Building regulations that require solar projects in the future could reduce the attractiveness of solar for property owners by mandating compliance.
- Insurance companies have only a short track record underwriting commercial solar projects and policies are likely to change as more claims data becomes available.
- Controlling the environmental attributes of a solar array could be beneficial where carbon emission reductions have been enacted or are being contemplated.
- Solar technologies continue to improve and have the potential to accelerate the obsolescence of existing solar installations.

In Chapters 11 through 15 we explored the key drivers of risk during the design and construction phases of solar projects. These were primarily aspects of the project that were fully under the project developer's control – such as calculating the allowable structural loads or installing walk mats to protect the roof surface from damage. This made decision-making straightforward: once a possible risk factor was identified a solution could be determined, priced, and implemented. But not all solar project risks have a solution that an engineer can design or a contractor can install. Some risks are outside of the direct control of the property owner or solar project sponsor. These risk factors tend to affect the industry as a whole and may in part be driven by regulations. This means that these risks are unlikely to have a resolution that can entirely mitigate the possible risk. On the other hand, the ability to identify and manage these risk factors effectively can become a competitive advantage for those who have a complete understanding of these risks and can make better-informed decisions.

This chapter focuses on the following areas where solar projects can face risks that are partially or completely outside of the direct control of the property owner or solar project sponsor:

- incentive program revisions;
- rooftop value;
- supply and demand for rooftops;
- valuation of buildings with solar projects;
- insurability;
- security;
- end-of-life system removal;
- building performance regulations.

These industry-wide and regulatory risks may result in unexpected benefits to property owners as often as they lead to negative outcomes. For example, regulations may increase the value of solar on commercial buildings by creating a market where verified carbon emission reductions from solar at commercial properties can be sold in a carbon market. It is only in the past decade that commercial property owners have been able to gain firsthand exposure with solar. As a result, part of the uncertainty is inevitable as the industry charts a new course with commercial solar projects. This means that a coordinated effort on the part of real estate and solar industry participants has the opportunity to help shape these risks into opportunities. With the right effort, well-informed property owners can identify the risk factors most likely to affect value for their properties and projects, and work to ensure that their investments remain profitable for years into the future.

Incentive program revisions

Good solar incentive programs seek to provide long-term certainty about the value of the financial support they provide in order to foster a healthy and stable marketplace. In nearly all markets, once incentives have been awarded they are permanently locked in for the projects that have received them. This is not always the case, however. In the past few years, some programs have been scaled back aggressively in the face of deep economic recessions and national budget shortfalls, including potential – and highly contested – reductions in payments to projects that have already been funded and are now operational.

Spain's solar program is an example of this situation, where incentives were essentially clawed back from projects after they had been awarded. For some solar installations, the feed-in-tariff (FiT) payment due to project sponsors under the national incentive program had been awarded. Based on this tariff rate developers decided to build, and investors agreed to finance solar projects. Then, in the face of government austerity due to a severe and protracted recession, tariff rates for already-awarded projects were cut. This extreme situation is not common, but as Spain illustrates, it can happen. If this does occur, it is largely outside the control of the solar project owner.

More generally, as the cost of renewable energy systems decline, financial incentives for solar are likely to be reduced or eventually disappear entirely. Today, many incentive programs already have steps that automatically lower their incentives based on achieving a pre-determined quantity of projects. Installations being planned at the cusp of one of these steps will face some risk that if the project is delayed or if the queue of projects fills up faster than anticipated, the incentive level could drop. This risk is outside the direct control of the project sponsor. The most common way to manage this risk is to underwrite the project assuming that the lower incentive level is available. If the higher level can be achieved then the project will see a financial windfall.

Rooftop value

Think of commercial rooftops as undeveloped land. In a real estate market where there is little demand for new buildings, land has correspondingly little value. In a market where there is demand for new properties, the value of land increases where it has the potential to support development. You can think of commercial rooftops in the same way when it comes to solar in a market that offers an attractive incentive program. In a market with little demand for solar, rooftops have little value, but in an emerging solar market the rooftops that can support solar projects may be in high demand.

Now consider that today's existing commercial buildings were not developed with rooftop solar project revenue in mind. As a result, in today's real estate market an arbitrage opportunity exists. Solar revenue can be generated on a commercial building where that income was never anticipated when the property was developed or acquired. An astute developer or buyer of commercial property could incorporate the value of future solar rooftop development into their underwriting in order to enhance their returns.

The trick to this is determining what value to place on solar. If a stable and well-established rooftop solar project market existed there would be comparable sales, or "comps," that could be used to infer the value of similar solar rooftops. But until the value of rooftops with solar potential becomes well-established, there will be a large amount of uncertainty as to what, if anything, is the revenue potential of the roof. This determination is heavily dependent on the stability of regulations and the effect they have on sustaining the market demand for solar projects on commercial properties. Changes in incentive programs could quickly alter the rooftop market and drive a change in value, either higher or lower. Even unexpected changes could affect rooftop value, such as change in structural code requirements that makes it more difficult for some roofs to support solar projects. These changes are largely outside the control of the property owner, but they can nonetheless affect the value of the rooftop and therefore the value of the building.

Supply and demand

A consideration that is closely related to solar value is supply and demand for commercial properties that are suitable hosts for solar projects. Some solar markets have quotas for commercial building projects, or a cap on the total capacity of solar installations their program can support. This effectively establishes a level of demand for host sites. Depending on the size of the market, the size of the solar program, and the financial attractiveness of projects, the demand for commercial building rooftop sites could be high or low. A small incentive program spread across all property types in a market may not drive much demand for host sites. Conversely, a large solar program that targets a portion of its incentive program toward commercial rooftop projects in the same market may spur significant demand for available sites.

The level of supply and demand may be difficult to determine with great accuracy, but an approximation can be obtained by comparing the square footage needed to meet the rooftop solar program requirements with the number of properties that are of suitable size and location to host solar facilities. This analysis will need to account for any competing projects that are able to apply to the same incentive program. Also account for operational limitations at commercial properties that are likely to keep some projects out of contention, such as insufficient roof structural capacity or unsuitable locations within the market. This can give you a sense of what the relative levels of rooftop supply are that may be available to meet solar incentive program demand.

You may be able to approximate supply and demand for solar sites with the right data and assumptions. First, determine the commercial building market's square footage that is eligible for the solar program. Subtract the square footage of sites that are too small or that are likely to pursue other programs that may exist for larger sites. Then reduce the eligible properties by a percentage that accounts for unsuitable properties due to location restrictions, insufficient roof size, roof age, structural limitations, and asset management considerations that may discourage owners from participating.

Let's walk through an example of a solar incentive program that has a cap of 250 megawatts and a duration of five years, targeting solar projects of 1,000 kW in capacity in a market that contains one billion square feet of commercial property. This gross market square footage needs to be adjusted to reflect the portion of the market that can be used for solar projects. These percentages are placeholders, but the numbers in Table 16.1 provide an illustration of how you can identify the supply of rooftop space available to meet demand.

You will need to adjust the percentages based on your knowledge of the property market and the details of the solar incentive program. Factor in any additional limiting factors that are relevant in the situation you are exploring. For example, a portion of the property market may be in a utility territory that does not participate in the solar incentive program.

In the example supply calculation we found that even though the total market is quite large, the net supply of suitable square footage is considerably smaller – less than 3 percent of the entire market. If each 1,000 kW solar project requires 200,000 square feet of gross roof area, our example appears to indicate that there is not sufficient rooftop supply to meet the program's demand. The program calls for 250 MW to be built over five years, or 50 million square feet per year. Compare that to the 27 million square feet available in a given year. You can see in this highly simplified example that demand could be expected to exceed supply. Based on this type of calculation you could infer what the potential value of rooftop space in this market could be worth today and in the future. With enough confidence in the numbers, it could influence your decision about when to enter the solar market.

Table 16.1 Market supply calculation

1,000,000,000	Gross commercial building market square feet
× 40 percent	Portion of market with a roof area sufficient to host a 1,000 kW solar array
× 90 percent	Properties that have structurally adequate roofs
× 30 percent	Properties where the owner is willing to participate
× 25 percent	Roofs replaced in each year; assumes 5 percent per year for five years
= 27	Million square feet of net available rooftop supply in each year of the program

From the example market rooftop supply calculation you can also see how even small changes to the numbers can change the bottom line outcome significantly. Currently there are few data to fill in the blanks with a high degree of precision. As a result there is uncertainty in determining supply and demand that cannot be controlled by the property owner. Making the best estimates you are able to will help you arrive at a number you have confidence in, but it cannot eliminate the margin of error associated with the calculation. As the supply and demand balance shifts in each market over time, it is likely to have an effect on the value of your property as a host for solar projects.

Valuation of buildings with solar

Rooftop solar facilities can provide a stable, low-risk source of revenue for the property owner. This revenue increases the net operating income of the property and contributes to higher valuations. This is true whether the project was developed under the direct ownership structure, the PPA structure or the lease structure. In all cases there is an increase in revenue attributable to the solar project. Where there is considerably less certainty is how the real estate industry will place value on the revenue. The property owner may be able to provide guidance in this regard, but ultimately the market will determine the value – how much a buyer will pay for a property that has solar.

Property owners understand that property income is subject to market cycles, vacancy, capital costs, and a host of other factors. The income from a solar project tends to be much more stable because it is driven by the value of the energy it produces. This does not vary with real estate market cycles and it is not affected by vacancy at the property aside from perhaps the PPA structure, and even then only in a situation of extreme prolonged occupancy. Because the drivers of solar revenue are different than those of the underlying property value, the solar revenue can be valued separately from the underlying building. This stability may suggest a lower cap rate and a higher valuation for the income from the facility compared to the building's operating income. When looking at all of the building's revenue sources, solar helps to diversify revenue streams and can reduce the overall volatility of the property's income. Exactly how much solar can be expected to deliver an increase in property value is not yet well established by the real estate market.

Quick tip

The long-term stability and diversification solar revenue provides to a property can result in a lower cap rate and higher valuation for solar revenue compared to traditional building tenant lease revenue.

The value of solar is likely to be enhanced if solar projects become standard features of commercial properties. In this case there will be more certainty that the building can count on solar revenue years into the future after the current equipment becomes obsolete. Without that certainty, there is a tendency to heavily discount the incremental value of solar revenue. This is because solar equipment wears out throughout its operational lifetime; solar modules degrade and electrical equipment approaches the end of its useful life. At some point it becomes time for the equipment to be removed from the building. You may recognize this – it can be thought of in terms similar to a ground lease, where the value of the lease declines as the number of years remaining becomes smaller. In the same way, as a solar project approaches the end of its life its value also declines.

If markets develop that ensure a steady demand for subsequent rooftop solar projects after the first one is removed, then the property owner will be able to count on an indefinite source of solar revenue. Value would be tied not only to the current solar facility, but also to the underlying value of the rooftop as a host site for future solar projects. This would enable solar project revenue to be underwritten as a recurring cash flow like tenant leases, rather than an amortizing revenue stream like a ground lease.

In valuation terms, the revenue of a solar facility that has been operating for only a few years may simply be capitalized at the same rate as the underlying property operating income since there are likely to be almost two decades of solar revenue yet to come. This may also be practical from an underwriting standpoint considering the relatively small increase that most solar projects will contribute to the overall value of a commercial building. As the solar facility ages, the revenue could be underwritten like a ground lease by calculating the net present value of the future solar revenue, assuming there is little or no residual value at the end of the term of the solar facility's lifespan. The cost of removing the solar array may also be factored into the present value as the system approaches the end of its useful life.

Solar projects add value to commercial properties today, but there is no standardized way to underwrite the revenue. This lack of clarity in real estate solar valuations leaves several justifiable ways to calculate project value, and therefore creates a financial risk for property owners pursuing solar today that their valuation differs from others in the real estate industry when the project is transacted.

End-of-system life

There are several unknowns when it comes to removing a solar array 20 or more years in the future. The first unknown is how long the array will operate productively. While the system output may be contracted for 20 years, the equipment could remain in fine working order for many years after that. Since solar modules now commonly carry 25-year performance warranties, there is a good chance the array could be expected to operate for at least that long. At the outset of the project, the array's condition decades later cannot be known with certainty, and its condition can greatly affect the decision of what to do with it in the future.

A second consideration for the end-of-system life is the roof and building underneath the solar array. The roof will be at least as old as the solar facility and there may be a need for repairs or replacement where the solar facility sits on the roof. Removing the solar array and re-installing it to facilitate a roof replacement may not be a justifiable expense with an old solar array. The cost of re-installing the array will likely require not only labor, but also replacing aged components such as wiring, racking, and conduits. Add to this the performance of the solar modules, which while perhaps still fully operational, will be significantly diminished, and this path may not make much economic sense.

A third consideration is that new and more efficient solar module technologies will exist when the solar array reaches the end of its operational life. They will also likely be much less expensive. So even though the existing installed solar modules may be operational, it may be more cost-effective to start over with a new roof and a new solar array using the latest technology. The economic analysis associated with determining the right time to replace a solar array cannot be determined today. This introduces some risk into the financial model used to value the solar project that cannot be controlled by the project owner. An array that will require replacement after 20 years presents a very different financial model than one that can remain in operation for 25 or even 30 years. The extent of improvements in efficiency and cost cannot be known today, so there is little ability to predict exactly how these changes will impact decision-making years into the future.

Regulations for solar array design and structural requirements are also likely to have changed between when the array was installed and when it is slated for removal. New regulations may dictate changes to the building or the solar project that make keeping an older system uneconomical. Conversely, new regulations may "grandfather in" old solar projects and provide an incentive for the property owner to keep them operating as long as possible. While regulations are likely to change, the extent and timing of the changes cannot be known today. This introduces some risk into the project that is outside the control of the project owner.

Building performance regulations

Commercial building codes are certain to evolve during the two decades or more that your solar project is operating. One nascent change that has begun to be considered in environmentally focused cities and regions are new regulations that would require new commercial developments to generate a portion of their energy from on-site renewable energy. In this scenario, three things are likely to happen. First, incentives to support the installation of solar projects are likely to diminish and disappear – why subsidize something that is required by law? Second, the differentiation that comes with a solar project today will diminish as more and more projects become hosts for renewable energy systems. And third, the space occupied by a mandatory solar array will reduce space available for a revenue-generating solar project such as a roof lease with the utility company.

This scenario may seem distant, but it is getting closer in some locations. Policy-makers in London, New York City, and elsewhere have been discussing requirements that commercial buildings produce a portion of their energy on-site. Some policy-makers are even considering net-zero energy building regulations. Net-zero energy buildings are characterized by their ability to produce as much energy as they consume throughout the course of a year. This is achieved through a combination of energy conservation strategies that reduce the building's energy needs and on-site renewable energy generation to meet any remaining energy requirements.

The emergence of on-site renewable energy generation regulations and net-zero energy buildings will shift the way solar projects are valued and it will impact the revenue-generating opportunity that exists today. For property owners considering solar projects in the near future this is not an immediate worry, but evolving regulations are likely to impact the value of solar on buildings as today's projects approach the end of their useful lives.

Quick tip

Policy-makers in environmentally progressive cities are considering rules that would require new commercial building developments to generate a portion of their energy needs from on-site renewables such as solar.

Carbon emissions

Carbon emissions regulations are a frequently discussed topic among environmental policy-makers around the world. The emergence of carbon emission reduction targets has the potential to increase the value of solar projects as sources of emission-free electricity. There are various ways a value can be placed on carbon emissions – carbon tax, carbon trading markets, renewable portfolio standards and the like. There is a great deal of uncertainty in most parts of the world as to how carbon emissions will be managed in the future, and whether regulations will even directly affect commercial property owners. This picture is likely to come into focus over many years, and the shape it takes will be largely outside of the property owner's control. The potential to attribute value to carbon emissions in the future could benefit solar project owners that are knowledgeable about solar. Those that are able to structure their projects in ways that ensure they control their project's environmental attributes stand to gain from the evolution of carbon emissions policies.

Emerging technologies

The solar industry's rapid growth is generating technological advancement that increases energy production, improves longevity, and reduces costs. This innovation is essential to make solar projects more cost-effective as renewable energy incentives continue to decline in markets around the world. The pace of change is steady, in part because many solar products are mature technologies, and in part because of the often-large capital investments required to scale-up new technologies to enter the market and become competitive. In a typical year, solar module efficiency increases by about 0.5–1 percent for commercially available products. While it is unlikely that a solar module installed today will suddenly become obsolete due to the arrival of a disruptive solar technology, it is almost certain that technologies will continue to improve and evolve.

Future developments in technology have little impact on the system you install today, but advancements may affect the speed at which your system becomes technologically obsolete. This evolution will deliver ever-improving solar solutions, at the risk of reducing the relative value of existing projects. Even if your system is operational in 20 years, new technologies may be sufficiently more productive and cost-effective that it makes more sense to install a new array than to continue operating an obsolete system. The inability to predict this advancement provides some inevitable uncertainty. Fortunately, the uncertainty will likely be a win–win of sorts: choosing to remain with an old-but-functioning system whose cost has been fully amortized versus moving to a new project that captures all the latest advancements at much lower costs.

Quick tip

Solar module efficiency has historically increased by about 0.5–1 percent per year. A 17 percent module today might be 18 percent efficient in two years' time.

Final thoughts

The systemic industry and policy risks identified in this chapter create uncertainty when attempting to understand long-term implications for solar projects and property owners seeking to host systems. Some are likely to fall in favor of increasing value and creating opportunity for property owners. Others could make solar less attractive by affecting revenue or reducing the feasibility of solar in the future. These risk factors are largely outside the direct control of property owners and the solar industry, but they can in many instances be influenced through education of policy-makers to help ensure they act in the long-term best interests of supporting clean energy and new sources of revenue for property owners.

Chapter 17

Warranties and insurance

Chapter summary

- Warranties protect against defective materials, poor workmanship, and out-of-specification performance.
- Insurance solutions help shield the project owner from the cost of solar project damage and limit lost solar revenue.
- Coverage of the underlying building should be reviewed any time solar is being considered to ensure that the project does not impact existing insurance policies.

Insurance is a risk management solution that property owners use to address risks and events outside their control. Insurance is a contract where an insurer receives premium payments in exchange for providing coverage of the insured in the event of a covered loss. This solution, which transfers financial risks to a third party, can be applied to various aspects of solar projects. The discussion on solar insurance focuses on the following areas:

- insurance related to solar equipment;
- insurance for solar project revenue;
- insurance related to the underlying building.

Solar equipment

Insurance solutions are available to protect the solar array owner from damage or loss that affects the solar array. These types of insurance protect against defective products and workmanship, accidental breakage, vandalism, theft, and other events that result in damage to the solar array. Insurance policies do not cover routine wear and tear, and they may not cover damage related to certain types of extreme weather events or terrorist attacks.

Product warranties

The first type of insurance for components of a solar project is the equipment warranty. A warranty states that if a product does not perform within a specified range, the manufacturer will commit to repairing or replacing the product. Warranties are available for all major project components, from solar modules and racking systems to inverters and transformers. Warranties are valid for a specified period of time. Warranties can also be voided if the terms and conditions of the warranty are not met. Failure to properly maintain an inverter or improperly installing a solar module could lead to a warranty becoming void.

Keep in mind that warranties are only as good as the company that stands behind them. In the past decade a number of companies in the solar industry have ceased operations. Along with the business closure comes the loss of the equipment owner's ability to seek warranty service. Unless the company was sold to another business that is still in operation, there is little that can be done to enforce a warranty from a defunct company. As you evaluate products for your solar project, keep in mind the health and long-term viability of the companies that supply the warranties.

Warranty insurance

Warranty insurance is a financial product that backstops an equipment warranty. In the event the manufacturer is unable to honor their warranty commitment, the insurance would step in to provide comparable coverage. This type of insurance is one way for new manufacturers entering the solar market to alleviate fears among potential buyers that the manufacturer will be around over the long term to service the equipment warranty. This is particularly relevant with products that are long-lived and have long-term warranties, such as solar modules. Many solar modules today come with 25-year performance warranties. Warranty insurance reduces the risk to the module purchaser that the manufacturer will be in existence to service a warranty a quarter-century later.

Performance guarantees

A performance guarantee is a type of product warranty that specifies a certain minimum level of output for a solar module. This guarantee helps to ensure that the product will reliably produce energy throughout its expected lifetime. This guarantee may specify that the module "will retain 90 percent of its minimum peak power output in the first ten years and 80 percent for the remaining term of the guarantee" – typically the next fifteen years.

Contractor insurance

Contractor insurance provides coverage for professional liability or errors and omissions coverage in the situation where there is a problem with the design or installation of the project. This could be due to poor workmanship, incorrect design, or negligence. This type of insurance can come into play if the system is underperforming due to a design or installation flaw, or if a problem with installation causes damage to the system.

For projects developed under the direct ownership project structure, these types of equipment insurance should be considered in order to be able to effectively manage the potential financial risk of developing and owning a solar project. For projects developed under the power purchase agreement (PPA) or lease structures, the third-party project owner will be responsible for insuring their own solar project. It is still prudent to ensure that the third-party solar project owner carries sufficient insurance so that any

Lessons from the field

What is the operational life of a solar facility?

The life span of a solar project can be defined in terms of how long the contract you've signed lasts, how long the equipment will function, or even how long the roof underneath the solar project will last.

Contracts

PPAs and roof leases are typically 20-year contracts, although they can range from 15 to 25 years. Extension options can be included in the contract.

Equipment

Solar modules are warranted for 20–25 years. The actual operational life can extend to 30 years or more in the right operating environment and with regular maintenance. Wiring and other electrical equipment will last the life of most projects when periodically inspected and properly maintained. Inverters generally carry up to 15-year warranties, although extended 20-year warranties are becoming more common. Other system components such as racking, conduits, and raceways will generally last indefinitely if inspected and maintained properly.

Roof

A roof under a solar array should be designed to offer 20 years of trouble-free life. Virtually any commercial roof system can be purchased with a 20-year warranty. If a roof needs replacement after the term of a PPA, it may not matter if the solar equipment is still in sound operating condition. The cost of removing and re-installing an old solar array would likely preclude the PPA provider from wanting to extend the PPA after the new roof is installed.

lack of coverage or other shortfall does not impact your use of the solar energy or your receipt of roof lease payments.

Solar revenue

There are several types of insurance available to protect the project owner in the event of a disruption in expected revenue streams associated with the solar project. The most common form of coverage is business interruption insurance, but insurance coverage can also be obtained for delays in the commercial operation date of the system and loss of tax credit revenue.

Business interruption insurance

Business interruption insurance is one of the most common types of protection against the loss of revenue from a solar project due to a disruption in its ability to produce energy. This type of policy replaces lost business income due to a covered incident during the operational life of the solar array.

System start-up delay insurance

In the event there is a delay in the commercial operations date of the project – when it begins to reliably produce power – system start-up delay insurance provides coverage for the potential loss of revenue. This type of coverage is usually triggered by equipment failure or some form of natural disaster or covered weather event.

Tax credit insurance

For projects that rely on the ability to use tax credit financing to fund construction, tax credit insurance provides protection against problems that can arise when qualifying for the credit. In the United States, where tax credits can account for one-third of project costs, it is important to capture the full value of this incentive. If a solar project ceases production it could be faced with a portion of the investment tax credit being clawed back by the government. This insurance can also provide coverage against changes in the tax treatment of the project that could affect the investment profile of the project. This type of insurance is more common on very large projects where the tax credit amount is essential to the viability of the solar entity financing the project.

Underlying building

For the property owner, their building will in almost all cases already have an insurance policy in place when solar is added to it. It is prudent to consult with your insurance provider to ensure that your current policy provides adequate coverage for your solar investment, particularly if the solar array has a high value relative to the underlying building. In this case, insurance premiums could rise if solar represents substantial value or if there is a special circumstance that introduces greater risk for the insurer. For solar projects developed under a PPA or roof lease structure, the underlying building owner should verify that their current insurance coverage for the existing property is not affected by the addition of a solar array on the premises.

Business interruption insurance should be considered to protect the property from a loss of revenue if the system goes offline for an extended period of time, even where the solar array is owned under a PPA and not directly controlled by the property owner. In this case, if the PPA provider is unable to deliver power to the property, the property owner will be unable to sell the power on to building occupants.

Roof warranties

A roof warranty is a form of protection that reduces the exposure of the property owner to costs due to a failure of the roof system to perform as the manufacturer intended. A roof warranty is not a panacea for eliminating roof maintenance costs. It will not eliminate your exposure to repair costs if the solar project causes damage to the roof system because this damage will in most cases not be covered by the roof warranty. Roofing manufacturer warranties largely restrict their warranty to manufacturing defects. Warranties exclude a broad range of conditions that the roof may experience with a solar project, such as:

- equipment resting on the roof;
- non-approved adhesives or solvents;
- excessive foot traffic;
- staging heavy or abrasive materials on the roof surface.

Manufacturer's warranties do not cover maintenance costs or damage caused by a solar array's contact with the roof. For example, if there is accelerated wear in the spot where the racking contacts the roof it could void the roof manufacturer's warranty. Some roof warranties are also non-transferrable unless consent of the manufacturer is received, which may require a roof inspection. This consent may be harder to get if the inspection reveals a roof extensively covered by solar equipment that the manufacturer was not aware of. This situation could come up if you are looking at purchasing a property that has solar on it, or if you decide to sell a building that has solar.

Check roof warranties for the total number of years their coverage is valid. Many standard roof warranties only cover the first ten years of roof life, even if the roof is expected to last longer than that. Purchasing an extended warranty is more costly and must be weighed against the likely value it could provide, given the exclusions previously noted. Extended warranties are typically available in five-year increments.

Keep these roof warranty considerations in mind when talking to solar companies and project developers. Some solar developers focus their efforts on roofs with long remaining warranties. Aside from giving an indication of roof age or expected life, a warranty provides limited value in determining a rooftop site's suitability to host a solar facility.

Quick tip

Undue focus on roof warranties by a solar developer may serve as a cautionary flag for property owners, indicating that the solar company is not focusing on other important due diligence efforts like roof condition assessments.

Roofing installers may provide a separate warranty against defects in workmanship, but this warranty seldom extends beyond one or two years. Repair or maintenance outside of this limited window becomes a cost borne by the property owner. Because of the short term of the warranty, it may be of limited value for roof work done to prepare for a solar project. Neither the manufacturer nor the installer's warranties provide protection against potential roof maintenance costs related to the solar facility. You can manage this roof maintenance risk in PPAs and roof leases by stipulating in contracts

---

Lessons from the field

Solar insurance standards

FM Global has developed criteria for evaluating the risk factors of solar project installations. They are also developing an FM Approval Standard that is expected to result in FM Approved solar products. They consider project risks in accordance with the following criteria:

- combustibility;
- wind uplift and securement;
- roof loading;
- roof drainage;
- other natural hazards resistance.

Standards such as those promulgated by FM Global are emerging as the solar industry grows, but they do not yet have widespread acceptance. You may find it valuable to educate your insurance provider about the scope of your project and measures that have been taken to mitigate risks. This may include ensuring compatibility and maintenance of solar array components, structural analysis, roof protection, and other proactive measures that benefit both the building and the solar project.

---

that roof damage or accelerated wear associated with the solar facility is the responsibility of the solar project owner. Contracts should also specify that the roof maintenance contractor must be acceptable to the property owner and the roofing manufacturer providing the warranty. This will help ensure that work is completed in a satisfactory manner without jeopardizing the roof warranty. Most solar companies are happy to use your current roof maintenance contractor since it provides continuity for the roof upkeep, helps keep the property owner happy, and saves them the trouble of finding one themselves.

Final thoughts

For any of the insurance coverages discussed in this chapter, recognize that there is a trade-off between the financial protection insurance provides and the cost to obtain it. You may decide that you are willing to take on certain risks because insurance coverage is not cost-effective. For example, you may be comfortable capturing the tax credits for your project. You may, on the other hand, be unwilling to take on business interruption risk in the event the solar array goes offline for an extended period of time. In this case you would either extend your existing building business interruption insurance or procure a separate policy specifically for the solar project. In many cases, the importance of insurance is related to the overall investment in solar and its relative financial impact on the property and the company sponsoring it. A large property owner may choose to self-insure a small solar project, while a small property owner pursuing a very large solar project may seek the protection that insurance can provide.

In the future, if insurers were to encounter greater losses at buildings that have solar arrays, they may modify policy coverage to limit their exposure to solar projects or increase premiums to price in the losses they are experiencing. In extreme cases, some insurers could decline to provide new coverage to buildings with solar on them. For example, in the aftermath of a major earthquake that saw significantly more building damage to properties that had rooftop solar arrays than those that did not, insurers may decide that they will no longer insure buildings with solar in that region. The only exceptions might be properties that have first undergone extensive seismic upgrades or other safeguards that comply with strict insurance underwriting standards.

Properties that are part of a larger portfolio where insurance is provided under a portfolio-based policy are less likely to be impacted by more stringent insurance underwriting requirements. This is because risks are spread out and underwritten across an entire portfolio, taking into account the overall profile of the assets as well as the individual properties. In this case, until the number of solar projects becomes meaningful to the portfolio it is unlikely to impact underwriting by insurance companies.

While solar property insurance is not new, its application to solar projects on commercial buildings continues to evolve as the insurance industry sees more and more systems installed. As the number of policies rises, insurance coverage and premiums are evolving as well. Policies are becoming more widely available and more cost-effective, particularly for solar projects that use high-quality components and industry best practices, and are installed in locations that do not present undue weather-related risks. Solar projects that use untested technologies or those from lower-tier manufacturers may have more difficulty finding cost-effective insurance coverage. Projects that are highly customized or installed in a non-typical manner may also have difficulty obtaining cost-effective insurance coverage. In cases where insurance coverage is not cost-effective or unavailable, the solar project owner may find it most feasible to self-insure the project.

Chapter 18

The future of commercial buildings and solar

Chapter summary

- There is untapped potential for commercial property owners to profitably host and develop solar projects.
- If a solar project creates value to justify the added risks by generating revenue, lowering operating costs or attracting tenants, it is generally a good decision.

Commercial property professionals have a unique opportunity to profitably support the growth of the renewable energy industry through solar projects at their properties. When you consider the many billions of square feet of commercial building rooftops, façades, and parking lots that receive sunshine every day around the world, it becomes apparent that there is enormous potential for the real estate industry to pursue solar projects that generate both revenue and clean energy. Planned and executed properly, these projects are profitable, add value to the underlying building, and deliver significant environmental benefits.

Throughout this handbook we identified the most relevant considerations for property owners interested in deploying solar projects in their commercial property portfolios. We provided guidelines to focus on what aspects of the projects are most important in order to manage risks effectively and ensure a successful outcome. Learning the lessons in this handbook allows you to plan and execute solar projects that provide years of low-risk revenue while producing a consistent source of clean energy that benefits you, your community, and the environment. Using the lessons described in this book to support informed decision-making for investment in solar allows you to validate the following assumption for your pursuit of solar that I highlighted in the Introduction to this handbook:

> If a solar project creates value to justify the added risks by generating revenue, lowering operating costs or attracting tenants, it's a good decision. Conversely, if the costs and benefits cannot be aligned, deploying solar may not be justified.

Following the guidelines in this handbook and combining them with the knowledge and perspective your real estate experience brings to the table will provide a valuable compass to use as you navigate solar projects. The material in this handbook is intended to complement your knowledge and strengthen it so you can make informed decisions that meet your needs based on your own opportunities and constraints. No single handbook or resource can describe every possible risk, project structure variation, or situation that you may encounter. With this handbook I endeavored to provide a thoughtful and methodical process that can be followed when evaluating and executing solar projects. By following this principle and complementing your own knowledge with the information provided in this handbook, you will be able to identify solar opportunities that exist in your portfolio and ensure that you are able to capitalize on them while managing your risks effectively. This is the secret to answering the question posed above, and to creating lasting economic value from solar projects.

Looking forward

Looking forward over the next few years and beyond, you can expect significant opportunity and innovation in nearly all aspects of the solar industry, and in its interaction with commercial real estate. The solar industry is evolving rapidly and with these changes come new opportunities to improve the quality of projects for commercial buildings. The cost to complete a commercial solar project has declined by more than 80 percent in the past decade and continues to drop. New solar markets are opening up around the world. Incentives have been scaled back in many markets around the world, and yet the growth of solar installations continues unabated. While I was writing this handbook, solar incentives in Japan were renewed and expanded, and China increased its long-term renewable energy targets and moved aggressively toward becoming the largest global solar market.

Utilities are coming to terms with the need to accommodate increasing amounts of distributed generation from solar arrays within their utility territory. The time, cost, and effort required to receive approval to interconnect a commercial solar project continues to drop as utilities upgrade their infrastructure and learn lessons about the effect of solar on their distribution network. As concepts like smart grids become realities, the ability to add even more solar on commercial buildings will become easier and more cost-effective. Property owners may even some day be encouraged by utilities to install solar as a solution to reduce peak energy demand in areas where the utility grid is most congested.

Contractors and other service providers are rapidly scaling up to meet the challenge of deploying solar on commercial buildings. Tradesmen are quickly gaining experience that enables them to deliver higher quality more quickly than ever. The considerations of working at occupied and constrained sites are not new, but the additional steps required to deliver a successful solar project take time to disseminate throughout the industry. These skills will continue to grow and foster innovation as more projects are developed.

Innovative financing solutions for solar projects that were not available a decade ago, such as the power purchase agreement are now commonplace in many markets. This financial innovation continues with solar leases and property-assessed clean energy (PACE) financing. There are likely other innovations to come that will further expand the solar opportunity for more properties.

The commercial real estate industry is also moving forward. I hear frequently from property owners and developers seeking information about how they can deploy solar on their portfolio. Industry panels and discussion groups routinely focus on the opportunities and pitfalls of adding solar. Renewable energy is taking on an increasingly prominent role in sustainable building certifications such as LEED, BREEAM, and others that many developers have incorporated into their specifications. Many property owners, particularly in sunny climates, are eager to make use of the abundant sunshine that they see every day. They are seeking additional sources of revenue to increase the value of their properties, but many are also genuinely intrigued by the opportunity to lessen the impact their properties have on the environment.

Growth and innovation in all these areas are taking place against the backdrop of a world that continues to see populations grow and consumption rise. This is occurring in ways that reinforce the importance of recognizing the value that the environment provides to society. Renewable energy projects, such as solar installed on commercial buildings, align with the need for this type of responsible growth and development. The contributions of the real estate industry toward meeting this challenge can be delivered in new and profitable ways that tread lightly on the environment. Solar projects on commercial buildings are poised to lead the way.

Part V

Case studies

The case studies in this part of the handbook illustrate several common issues that arise when analyzing a real estate portfolio, negotiating contracts, and deploying solar projects. Case studies include:

* selecting buildings from a property portfolio;
* when a tenant requests solar;
* choosing a project structure;
* selecting a racking system;
* comparing project structures for a build-to-suit.

These examples are taken from my experience working with commercial property portfolios in markets that have utility-managed solar incentive programs, primarily in North America. The issues that came up may be familiar to you if you have spent much time exploring solar for your properties. You will see in the case studies my approach to seeking solar opportunities. This process requires solar knowledge and experience that is used to guide project stakeholders toward an understanding of their options and opportunities. This includes educating project stakeholders so they can apply their own real estate experience to understand opportunities and find the best solution that meets their needs.

Chapter 19

Selecting buildings from a property portfolio

In 2008 the solar market in California was heating up. Electric utilities had launched well-funded, multi-year solar incentive programs that were generating significant interest from property owners and solar companies. These programs enabled solar projects of a range of sizes to be developed on commercial buildings across the state.

The statewide program covered many of the major commercial property markets, and the potential economies of scale with larger solar arrays led many property owners and solar companies to pursue projects on multiple buildings. The real estate company where I worked owned a portfolio distributed throughout northern and southern California. Executives at the company were interested in finding out if there were revenue opportunities that would justify deploying solar on their buildings.

We began due diligence on several fronts. We reviewed the basic criteria for the solar programs in both northern and southern California through conversations with utility representatives, online program descriptions, and discussions with policy-makers. We established a set of priorities that we wanted to achieve by pursuing solar. Our list of priorities included:

- generate incremental recurring revenue;
- offset, or at least minimize, upfront due diligence costs;
- minimize risks of negative impacts to tenants;
- avoid negative impacts to the underlying properties;
- establish a scalable program that could be replicated efficiently for additional properties;
- gain a complete understanding of the solar market and solar project financial structures;
- enhance our sustainability program reputation.

Our internal due diligence process focused on the following areas:

- Choose a subset of properties that are likely candidates for solar.
- Identify key go and no-go issues related to property lifecycle, asset management plans, and tenants.
- Review the property list with key stakeholders to secure their support.
- Perform physical due diligence for the candidate properties.
- Decide how we were going to develop the solar projects, i.e., would we do it in-house, would we share responsibilities with a third-party solar developer, or would we outsource the entire process?

We created a list of properties that generally fell within the criteria of the program. This included properties in two different utility territories. We held conversations with several solar service providers, primarily solar project development companies that were active in California and nationally.

Our external due diligence process focused on the following areas:

- understanding how the solar market in California compares to other markets, in order to know if California was really the right place to concentrate our efforts;
- identifying long-term solar industry trends, so we would know the best time to pursue solar;
- identifying the solar service providers best-suited to support our projects.

To assess the property portfolio, we divided the properties into two groups, one for the utility territory in northern California and the other for southern California. We gathered basic property information including the square footage of roof area, property ownership entity, expected roof replacement date, and roof type.

We reviewed tenant leases at the properties, and noted where tenants were responsible for the maintenance and upkeep of the entire property, including the roof. We considered whether there were any building tenants that were unlikely to support solar. We noted that there were several types of property ownership in the portfolio – some were owned by an investment fund that we managed, others were fully owned by our company, and others were owned by a joint venture (JV). In each of these cases, we tentatively crossed off our list the properties that appeared to have complications – such as properties owned through a JV that would require partner approval throughout the solar process. We did this because we were looking for an easy win and we wanted to avoid properties that presented obvious obstacles that other potential sites did not.

We shared this information with several solar service providers – primarily solar development companies – active in the state. These service providers provided their own layer of feasibility analyses. Their criteria identified the sites that were the right size to qualify for the utility's incentive program based on a schematic layout of solar modules.

The solar companies looked at the location of the properties and highlighted sites that were located close to areas that had sufficient electric grid capacity, because this would eventually make it easier to get approval from the utility to connect to the utility grid. We jointly decided that rooftops that were in the middle years of their lifespan were not good candidates because of the likelihood of a roof replacement during the life of the solar array. Several properties were removed from the list as a result. At the end of this process, we determined that there were sufficient properties that had passed the initial screening process to continue pursuing potential solar opportunities.

As a ranked list of properties was being assembled, we were discussing potential models for deploying solar projects internally and with solar development companies. We were also refining our own financial models in order to understand what the financial impacts would be for various project structures. We were considering the direct ownership structure, as well as the PPA structure and the roof lease structure. Any projects would need to provide a meaningful financial benefit for the underlying property and our company.

We refined our financial model based on knowledge from prior research and from solar estimates provided by the solar companies. These estimates included both project costs as well as system production estimates. This allowed us to identify the revenue that the potential sites could provide under each of the three project structures. We also decided that in any scenario the underlying property would need to receive the majority, if not all, of the revenue from the solar array. We made the decision to maximize value for the property since it would bear the greatest long-term solar obligation as the host. This was also needed to get the individual fund managers that controlled the properties to support the projects.

We found that the development structure offered significantly higher revenue potential from development fees while also paying the underlying property for the use of rooftop space. The trade-off was that we would have full responsibility for developing and operating a solar array – something that we had not done before. This path would also have to compete for capital with other real estate deals within the company.

The PPA structure had the advantage of allowing us to outsource the solar development risks to companies that had more solar expertise than our development teams. The trade-off was that the revenue potential was reduced to a fraction of what it would be under the development model. The fatal flaw in the PPA structure was that as the property owner we would have to take on the risk of purchasing the solar power and then re-selling it to our tenants. This was ultimately a solution we discarded because we did not want to take on this task or its associated risk.

In the utility territory where the buildings were located, a roof lease structure was also a viable option, albeit under a separate utility program. This structure provided a similar financial benefit to the PPA structure, but without the complexity of re-selling power to building tenants. The lease also carried less risk than the development structure, an important consideration when we were just beginning down the solar path. The lease was also a contract that our local staff felt comfortable with since they handled leasing on a regular basis. For these reasons, we decided to pursue the roof lease model for our initial solar projects.

Once we decided on the project structure, we signed an option agreement with a solar development company that would provide turnkey project development and operations. The option agreement provided them with security that both parties were committed to jointly pursuing the project. A version of this option agreement was also required to be included in the application to secure a place in the utility incentive program queue. At this point several sites were dropped from the list because they were too small to qualify for the roof lease-based incentive program (they would have qualified for the direct ownership and PPA-based incentive program in this utility territory). The option payments covered our due diligence costs up to that point in the process.

Because the buildings in the portfolio varied from a few years old to almost 50 years old, the solar company conducted structural due diligence for each site to identify rooftops that had a limited ability to carry additional weight. In conjunction with this analysis they refined their solar array design so they would know exactly how many kilowatts each site could safely support. During this process we conducted a peer review of the structural analysis to ensure the designs were not going to have an unwanted impact on our properties. The analysis found one property that had severe weight constraints and two additional properties that had borderline weight constraints. The most weight-constrained building was removed from the list. For the other two sites, the solar company made two changes. First, they switched to a racking system that was anchored to the roof with minimal ballast in order to reduce weight. They also reduced the size of the solar array and relocated it away from areas of the roof where the structure was unable to support the additional weight.

From start to finish, the due diligence process reduced the number of suitable solar sites in our initial portfolio by more than 50 percent. For the sites that remained we had confidence they were strong candidates for the successful deployment of solar projects that could be built as designed, would not have negative impacts on the underlying property, and would provide reliable revenue for many years.

Chapter 20

When a tenant requests solar

The company I worked for was negotiating a lease with a tenant for space within one of our business parks. The company produced a product used by the renewable energy industry, so it was important for their own business image to utilize solar energy. This case study illustrates how the local leasing team worked with the tenant to add an amendment to the landlord's standard lease to make it possible for the tenant to install their own solar array.

I received a call from the local leasing manager at our company, and he explained that a tenant was interested in solar. They wanted to deploy their own solar array on the building to be consistent with their corporate goals as a renewable energy product company. The challenges we identified were as follows:

- The proposed lease had a term of five years, with two five-year extension options.
- The tenant was leasing in a multi-tenant building.
- We had to ensure the roof was in good enough condition to support solar.
- The tenant was a fairly new company and had limited credit.
- We had to use a design that worked for both tenant and landlord.

To address these issues, we convened a series of meetings that included the following stakeholders:

- landlord's leasing manager;
- landlord's property manager;
- landlord's construction manager;
- landlord's solar advisor;
- tenant's leasing and facility representatives;
- tenant's solar contractor.

Five-year lease term

As previously noted, the typical contractual term for solar projects is 20 years. At the short end you may see 15-year terms, but not five, which was the term of this tenant's proposed lease. This meant that the tenant, the landlord, or both parties would take on the risk of amortizing the solar project costs over an unusually short period of time. The landlord did not want to take on this risk because they did not want to be responsible for the solar array. The tenant wanted to defray solar costs by incorporating the project into tenant improvements specified in the lease contract, and by having the landlord take over the solar array if the tenant decided not to renew after five years. This would save the tenant the cost of removing the array from the building.

The landlord identified several possible solutions. These included:

- refusing the tenant's request to install solar;
- bringing in a third-party PPA provider to own and operate the solar array and sell power to the tenant;
- amortizing the cost of the array into the lease;
- making the tenant responsible for all project costs including removal of the solar array at the end of the lease term.

The landlord decided that refusing the tenant's request would not be a productive way to build a relationship with this tenant. But before the landlord decided against saying "no," we did a quick check to verify whether the building was a candidate for any larger, more lucrative distributed generation projects in that market. It was not. If it had been, they might have reserved the roof for their own solar program rather than allowing the tenant to use the space. We also commissioned a quick structural review to verify that adding solar would be unlikely to be limited by the roof structural capacity. The building passed the initial check by the landlord's engineer.

We considered bringing in a third-party PPA provider to develop, own, and operate the solar array, but this was quickly dismissed. First, the tenant wanted to own the solar array, and second, due to the five-year lease term and the limited credit quality of the tenant, a PPA would not have been viable.

We next considered providing a turnkey solar solution and amortizing the cost of the solar array into the lease. The landlord had concerns that if the cost to install the solar array were amortized into the rental payments it would create a very large increase due to the short five-year recovery period. The leasing manager decided that adding this additional cost could put enough of a burden on the tenant that it could eventually jeopardize the tenant's ability to pay their base rent.

This issue was resolved when the tenant agreed to purchase the solar array with cash on hand. The lease specified that the tenant would remove the solar array at the end of the lease term and repair any roof damage associated with the project. The lease clause that addressed this looked like this:

> Upon the expiration of the Lease Term (as may be extended), Tenant, at Tenant's expense, shall remove the Tenant Solar System and all associated improvements and equipment and Tenant, at Tenant's expense, shall repair any damage to the Building or Project caused by the removal of the Tenant Solar System and associated improvements and equipment.

Leasing in a multi-tenant building

The tenant was one of several tenants in a building within a larger business park. The landlord's standard lease stipulated that maintenance of the exterior areas of the building, including the roof, were controlled and maintained by the landlord. This was the basis for the landlord's ability to decide whether to allow solar. A multi-tenant building meant two key topics had to be addressed:

1   ensuring that any solar-related costs were not charged to other tenants;
2   ensuring that both landlord and other tenants maintained rights to install rooftop equipment.

We addressed the first point by including language in the lease that made the tenant responsible for additional maintenance costs related to the array, such as new leaks, added inspections, or other roof wear due to foot traffic on the roof related to the solar project. The second point was addressed by limiting the array to the area above the tenant's space. The lease language looked like this:

> (i) Notwithstanding anything contained in the Lease to the contrary, Tenant shall be responsible for 100% of any costs incurred by Landlord for the repair of any damage to any portion of the Project caused by Tenant's installation, maintenance, operation, and removal of the Tenant Solar System. Notwithstanding anything contained in the Lease (as modified by this Amendment) to the contrary, in the event any repairs to the Building which are structural in nature are required as a result of the Tenant Solar System, such repairs shall be made by Landlord but, to the extent same are required as a result of the Tenant Solar System, Tenant shall reimburse Landlord for the reasonable cost thereof no later than thirty (30) days following Tenant's receipt of an invoice for such repairs.

> (ii) Landlord shall have the right to install rooftop equipment, including but not limited to other solar/photovoltaic systems, on the roof of the Building. Tenant hereby agrees that Landlord shall have the right to lease the Remaining Roof to another party who shall have the right to install, maintain, and operate within the area of the Remaining Roof only an electric generating, photovoltaic/solar panel system and any ancillary equipment associated with such system.

Age of the roof

The building was built in the early 1990s, and the roof had been replaced in 2011. The roof condition report noted that the roof had an expected remaining roof life estimated at 19 years. This meant that the roof would likely last for the useful life of the solar array. This meant that proactive roof replacement or coating was not required.

Limited credit tenant

The landlord acknowledged that the relatively small size of the tenant and the volatility of the renewable energy industry meant that if the tenant's industry slowed they could face financial trouble. For this reason the landlord was unwilling to take on costs associated with the tenant's solar project since it had a relatively high price tag. Since solar was a "non-standard" building feature in that market, the landlord did not think it could reliably recoup associated costs from future tenants if the building became vacant.

In early discussions, the tenant wanted to know if the landlord would share in the project cost or include it in the other tenant improvements planned for the space, costs that would be amortized into the recoverable costs in the tenant's lease. One suggestion was to consider extending the base lease term to ten years or more. This would have allowed costs to be amortized over a longer period, but there would have been greater risk that if the tenant ever became insolvent, the cost for non-standard improvements might not have been recoverable. The landlord decided that an initial ten-year lease term was not a good solution, but did want to provide a way for the tenant to get full use of the solar array. They negotiated two five-year extension options to the lease to address this. For their part, the tenant also did not want to sign up for a ten-year lease term so this topic was relatively easy to come to an agreement on. When the tenant decided to purchase the solar array the issue largely resolved itself.

Finding a suitable design solution

The tenant initially wanted a solar array that covered the roof of the entire building. The landlord limited that to the area above their leased space to avoid raising concerns with other tenants. This also provided the flexibility for other tenants to add rooftop equipment or pursue solar in the future, on the assumption that this project might pique their interest. We reviewed the proposed solar project design, and required the tenant to perform a structural analysis of the building to ensure that the roof would support the added weight. Based on our recommendation, they hired a firm the landlord had worked with extensively. This simplified the process of getting the landlord comfortable that there would be no adverse effects on the building.

Since the tenant was purchasing the solar array they selected the contractor that installed the system. The permitting and construction process for solar was independent from other tenant improvements. The tenant's solar contractor did coordinate with the landlord's construction manager to prevent conflicts with other building improvements being installed for the tenant.

One area where the landlord's construction manager focused their attention was on the way the racking system connected to the roof of the building. The initial proposal specified a residential-type metal flashing surrounding stanchions anchored to the roof structure. This solution would not have met the roof system manufacturer's warranty requirements, plus it did not provide consistent anchorage into the roof structure. The detail we settled on required the installation of wood sleepers anchored to the roof structural members and embedded in the roofing system. This provided an easy-to-install way to anchor the racking system that complied with the roof system's warranty.

The negotiation and planning for the solar project took place over the course of about one month, happening alongside other lease negotiations. In the end, the tenant's request to install solar was resolved to the satisfaction of both the landlord and tenant by getting the right people in the discussion from the start, and by working together to quickly resolve challenges that were identified by both parties.

Chapter 21

Choosing a project structure

In 2009 the solar market in Ontario, Canada emerged as a potentially lucrative location to pursue projects when the local electric utility launched a feed-in tariff (FiT) program that supported rooftop projects on commercial buildings. Based on preliminary research of the market we decided that it could be a good fit with our property portfolio. I produced a report that highlighted the opportunity and presented it to the heads of that region. A key element of the memo and subsequent discussions was the process of choosing which solar model to pursue for the projects in the Ontario market. The executive summary of the memo read, in part:

Executive Summary

This memo provides information on a rooftop solar photovoltaic (PV) opportunity emerging in Ontario, Canada. Ontario has created a rooftop solar program that would allow us to monetize our existing rooftops. We can generate revenue from rooftops in two ways: 1) Leasing rooftops to solar power developers and 2) Developing solar projects and selling electricity to the Ontario Power Authority directly. The risk and revenue opportunity vary based on the option selected.

This presents a type of arbitrage opportunity – to develop and monetize rooftops with income-producing solar projects that were not possible when the property was first developed. We believe this revenue opportunity in Ontario can be significant when compared to in-place rents in the underlying properties. We believe that our portfolio is in a highly competitive position to capitalize on this emerging program.

The Ontario market enabled each of the three solar project structures discussed in this book. As a property owner, you could develop and own a solar project yourself. You could sign a power purchase agreement (PPA) with a third-party solar company that would develop, own, and operate the solar project and sell you the electricity, and you could also lease rooftop space to a third-party developer that would sell power directly to the utility.

The development structure and the lease structure quickly emerged as the most viable options for projects on commercial buildings under the FiT program. The FiT was not available for projects such as PPAs – projects that are designed to be installed "behind the meter" – to offset on-site energy needs. While this type of project could be built and installed in this market, it would not be eligible for the FiT revenue and it was therefore deemed infeasible for commercial applications like the ones being considered.

With the PPA structure out of the picture, these projects were left with the direct ownership structure and the lease structure. The memo to the market managers summarized the general economics of each model. Note that the excerpt (Table 21.1) is written from the perspective of the property owner and developer.

Indicative economics

The memo provided a range of metrics based on pro-forma modeling of a 1,000 kW rooftop solar array, illustrating the key differences between the two structures. This side-by-side summary provides a way to understand how the two structures compare. If the PPA structure had been a viable option it could have been included in the same type of analysis.

The memo proceeded to outline the framework for pursuing opportunities in this market, which was slated to launch in a few months. Following discussions with the regional leadership, they asked questions that centered on the following issues:

- whether other property owners were pursuing solar, and with which model(s);
- how certain the economics were for each model;
- responsibility for maintenance and the associated costs in each model;
- whether or not the building would receive the solar energy in either case.

Table 21.1 Excerpt from an investment structure memo

Roof lease	Developer/owner
Indicative economics	
• $0 capital cost	• $4,600,000 capital cost
• $0 operational cost	• $30,000/Yr. operational cost
• Annual revenue of $0.15 per square foot	• Annual revenue of $1.40 to $1.60 per square foot
• (n/a) IRR (no investment capital contribution)	• 7.5% IRR on 20 year hold *
• Solar NOI contribution: four percent increase in underlying property NOI	• 11.2% IRR on sale in year 7 **
	• Assumes no debt is used
	• Solar NOI contribution: 28 percent increase in underlying property NOI
• A one-time $10,000 expense degrades IRR by 60 bps	• Susceptible to system downtime
	– Each month of lost electricity sales revenue degrades IRR by 10 bps
• Not susceptible to system downtime	• Susceptible to large one-time capital costs
• Not susceptible to large one-time capital costs	– A re-roof event during operational life degrades IRR by 80 to 100 bps
Other considerations	
• Solar development partner needed for regulatory compliance in either scenario	
• Accelerated roof wear, roof leak potential	
• Tenant impact from construction and maintenance	
• Suitable roofs must be <3 years old	

Notes
* 20-year hold assumes $0 residual value for PV system.
** Sale assumes 9 percent cap rate.

Subsequent due diligence and continued dialogue with the team in that region produced answers to these questions. We found out that property owners were pursuing both the lease structure and the development structure, based in part on whether they had any in-house expertise in development. We provided regional leadership with additional breakdown on the capital flows of the projects based on a seven-year hold period, at which time we assumed the underlying property would be sold, and along with it the solar asset.

This comparison provided several metrics that were used to gauge the impact of solar project structure economic performance. We compared capital costs for each structure. Net operating income (NOI) was then compared, as was the computed internal rate of return. Each project structure was looked at in the context of its impact on the underlying building's financial performance, not just as a stand-alone project. In particular, this analysis calculated the impact on property-level NOI. The difference between these two examples is significant – 4 percent versus 28 percent. We discussed how the financial, development, and operational risks differed significantly between the two structures.

For the property owner, the ability to put the solar income stream into context was a valuable way to demonstrate the tangible impact solar could have on revenues in the portfolio. In either case, there would have been a noticeable change in NOI for the property. For the lease model it would have been incremental, and for the development model it would have been significantly greater.

Other considerations that came up in the discussions on choosing a project structure included the following:

- availability and expertise of in-house development staff to oversee solar projects;
- use of capital for real estate investments vs. solar investments;
- relative attractiveness of different levels of income compared to risks;
- length of time the building would be owned.

We also performed sensitivity analysis around many of the inputs of our financial models. As noted, the results of this analysis were examined in the context of their impact on the underlying property. Several of these included modeling changes to:

- annual revenue;
- sale year;
- cap rates;
- residual values;
- operational life of the system.

By performing a side-by-side comparison of the viable project structures in this market, we were able to provide the information necessary for the regional leadership to understand the solar opportunity. We were able to put it into the context of the properties where solar would be deployed so they could see how it would affect the underlying asset. This information was ultimately used to make informed decisions about the pursuit of solar projects in this market.

Chapter 22

Selecting a racking system

Selecting the right racking system is essential to a solar project that minimizes impacts to the underlying building. This choice often focuses on the weight of the system as much as the particular manufacturer's design. In the end, the choice often comes down to whether the array is physically anchored to the roof or if it is held onto the roof with ballast.

A property owner hired a solar contractor to develop a 200 kW rooftop solar array on an office building. The contractor's initial design for the project specified a fully attached, or mechanically fastened, racking system. This design would have required more than 1,200 rooftop penetrations to adequately anchor the system to the roof. The property owner was uncomfortable with this many penetrations because of concerns about potential leaks. They requested alternatives that would reduce the number of penetrations.

The solar system designer had previously reviewed the structural analysis of the underlying building's roof and determined the maximum allowable additional solar loads the building could support. Because this was a historic building, the roof had little spare structural capacity. This was particularly acute in certain areas where the roof had longer spans. This meant that one alternative – a fully ballasted racking system – would add too much weight to portions of the roof. Spreading the array out across the roof to reduce its point loads was deemed unfeasible because little spare roof space was available, so this option was discarded. As an alternative, the designer specified a hybrid system that used both ballast and mechanical fasteners.

The designer had several racking systems to choose from that could be anchored to the roof and also ballasted. These products also had several angles at which the modules could be placed to capture more sunlight. The options explored included zero degrees, five degrees, and ten degrees. The property owner decided that the ten-degree tilt would impact the appearance of the historic façade. The zero-degree tilt reduced system output and also would have required more frequent cleaning. That left the five-degree tilt as the preferred solution.

The racking system choices had a different solution for supporting the solar modules. One system provided a support under each corner of a module. While corners where several modules came together could share one support, this still meant there were supports every six feet across the roof, at the joint between each module. This was a concern for the property owner for roof maintenance and accessibility. This layout also meant that some supports would not fall directly on top of the underlying roof structure.

This led to a switch to a rail-based racking system. This type of system used a metal frame that supported the solar modules. The frame spanned the width of 2–3 modules between supports. This change allowed the designer to reduce the number of penetrations to fewer than 70, while remaining within the weight limits of the roof. This design interspersed ballast between the mechanically fastened points of anchorage. The location of the supports could also be adjusted to hit the underlying structure while still providing adequate support for the rails.

Because there were fewer fasteners in this new design, each one needed to be firmly anchored to the structural frame of the roof. They could not simply be anchored into the old roof decking. The designer specified that wooden blocking would be used to support the racking system anchors and connect them to the structural members underneath. This solution provided a firm anchorage for the racking system, while also enabling the supports to be installed flush to the roof surface so they could be flashed properly to prevent leaks.

This hybrid racking design addressed the concerns of the property owner and provided a cost-effective racking solution for the project. By exploring a combination of racking design solutions, alternative anchoring solutions, and careful placement of fasteners, the project was able to meet the weight limits of the historic host building while minimizing the number of fasteners needed to anchor the array to the roof.

Chapter 23

Project structures for a build-to-suit development

A development manager in New Jersey had questions about how to address a request from a build-to-suit customer that requested he incorporate solar into the development project. Because this project was being built to the customer's specifications, the costs of the project would be incorporated into the deal and amortized into the lease rate. This would also include the solar array costs. The customer wanted to explore adding solar to reduce operating costs and meet corporate sustainability goals.

The objectives of the development manager were to meet the customer's expectations and deliver the building they requested at the lowest cost. He wanted to be sure that the additional cost for the solar array was accounted for correctly in the development financial analysis, and that the cost could be translated into the rent calculation to adequately compensate for the additional capital investment and risk.

The building was going to be a logistics facility for a global shipping company, and the approximately 1,000 kW solar array would be installed on the building's roof. In this market and for this development there were three scenarios for pursuing solar:

1   The developer owns the solar facility and contracts for construction and operations and management (O&M).
2   The developer leases the roof to a solar company that sells power to the customer under a power purchase agreement (PPA).
3   The customer develops their own solar array after the development project is completed.

In the first scenario the developer takes on responsibility for providing a turnkey solution for the customer. In the second scenario, the developer brings in a third-party solar company, and operates as a middleman in a solar power sales transaction. In the third scenario, the solar is split apart from the development project and handled independently as a separate project.

For each scenario, we outlined the pros and cons for the developer, excerpted below:

Scenario 1: developer owns solar facility

Benefits

- Highest revenue for developer.
- Flexibility to manage the solar asset (requires six-year lock-out for tax purposes).
- Control of solar technology choice.

Challenges

- Capital cost of approximately $5 million.
- Development, financial, and operational risks.
- Need to monetize tax equity via third-party investor.
- Having a tax equity investor may complicate sale of the asset.
- sREC credits cannot be transferred or sold to customer.
- Learning curve (first solar project by developer).
- Schedule to execute solar transaction may lag build-to-suit project.
- May need to revisit deal summary approved by Investment Committee.

Scenario 2: developer leases rooftop to solar company

Benefits

- No capital investment; minimal development, finance, and operational risks.
- Few management or sale-of-asset restrictions.
- Turnkey solar project can be implemented into existing build-to-suit project schedule.

Challenges

- Low annual revenue compared to owning solar array.
- Limited operational control over solar facility.
- sREC credits cannot be transferred to customer.

Scenario 3: customer constructs solar after development is complete

Benefits

- No capital investment by developer.
- Separates development transaction from solar transaction.
- Eliminates risk of delays to development schedule due to solar.
- Enables customer to procure solar when and how they want.
- Capturing tax credit incentives is handled by the customer.
- sRECs can be retained by the customer.

Challenges

- Developer gives up control over the solar aspects of the project.
- Eliminates opportunity for higher rental rate due to amortized solar costs.
- No revenue opportunity related to solar from leasing roof to third-party solar company.

Scenario 1 required the developer to take on certain risks to deliver the project in parallel with the build-to-suit project, but offered the greatest revenue potential since the development costs would be recaptured through a higher rental rate and a commensurate increase in the development fee. The sale of electricity was not factored in because the customer would take full operational control of the solar array once they moved into the facility, and be responsible for the production and use of the solar electricity.

Scenario 2 had no capital cost for the developer, so they would not have to pass on an increase in the rental rate to the customer. This scenario offered an opportunity for a modest amount of revenue from leasing the rooftop to a third-party solar developer. Scenario 2 would provide a turnkey solar project delivery solution for the developer that would eliminate potential tax credit capture challenges, such as bringing a tax equity investor into the deal.

Scenario 3 effectively took the developer out of the solar picture entirely, and handed over responsibility for the solar project to the customer. This was in many ways the cleanest solution since it kept the build-to-suit transaction and the solar transaction separate, but it meant that the developer was also relinquishing any control or revenue opportunity related to solar.

The responsibilities for the various aspects of the solar projects were summarized into tables. These were discussed with the developer in order to ensure that there was a full understanding of the requirements for each one. The tables on the following page describe these responsibilities, where the black-marked items indicate primary responsibility and the gray-marked items represent secondary responsibility.

Based on this analysis and other supporting financial analysis, the developer decided that they were not in the best position to develop the solar array as part of the build-to-suit project. Either the second or third scenario was the preferred solution. The developer went back to the customer and explained the benefits and challenges for each of the three options from their point of view. In the meantime, the customer had been reviewing their own goals for the solar project and performing their own financial analysis.

The customer concluded that the incremental cost in monthly rent for the developer to build the solar array was higher than their corporate real estate group wanted to see. They also recognized that developing the solar project after the development was complete would simplify each transaction by keeping them separate. And finally, the customer determined that they wanted to retain the sRECs associated with the solar electricity so they could count them toward their corporate carbon reduction commitments.

This led the customer to decide to separate the solar transaction from the build-to-suit project and develop it after the building was complete. This solution was embraced only after both sides had reviewed the analysis of the three options and had considered other subjective factors. In the end, both the customer and the developer agreed that this solution was the best way to integrate solar into the building.

Scenario 1 responsibility matrix: developer owns solar facility

	Developer	Customer	Solar company
Capital cost	■		
Construction and operation	■		
Tax equity revenue capture	■		
sREC revenue capture	■		
Power sales	■		
Vacancy/power sales risk	■		
Structure/roof upgrade costs	▨	■	
Roof maintenance	▨	■	
Solar technology choice	■		
System removal at end-of-life	■		
Energy sales/grid compliance	■		

Scenario 2 responsibility matrix: developer leases rooftop to solar company

	Developer	Customer	Solar company
Capital cost			■
Construction and operation			■
Tax equity revenue capture			■
sREC revenue capture			■
Power sales	▨		■
Vacancy/power sales risk	▨		■
Structure/roof upgrade costs	▨	■	
Roof maintenance	▨	■	▨
Solar technology choice			■
System removal at end-of-life			■
Energy sales/grid compliance			■

Scenario 3 responsibility matrix: customer constructs solar after development is complete

	Developer	Customer	Solar company
Capital cost		■	
Construction and operation		■	
Tax equity revenue capture		■	
sREC revenue capture		■	
Power sales		■	
Vacancy/power sales risk		■	
Structure/roof upgrade costs		■	
Roof maintenance		■	
Solar technology choice		■	
System removal at end-of-life		■	
Energy sales/grid compliance		■	

Appendix

Solar resources

Many solar company websites offer easy-to-understand illustrations of the process of converting sunlight to electricity. These can be found by typing "how solar works" into a search engine. Additional reference information can be found on the following websites.

- US Department of Energy, Energy Efficiency and Renewable Energy: www1.eere. energy.gov/solar_glossary.html
- Solarbuzz glossary: www.solarbuzz.com/Consumer/glossary.htm
- A primer on solar energy and solar resources is provided by the Union of Concerned Scientists: www.ucsusa.org/clean_energy/technology_and_impacts/energy_technologies/how-solar-energy-works.html

There are also many books and online resources for the engineering-minded reader or do-it-yourselfer interested in the nuts and bolts of solar array design. These are generally focused on smaller-scale solar projects than you are likely to encounter on commercial buildings. The basic definitions of solar electric systems for both commercial and residential solar projects can be found in books such as the following:

- *Photovoltaics: Design and Installation Manual*, by Solar Energy International (New Society Publishers, 2004).
- *Power from the Sun: A Practical Guide to Solar Electricity*, by Dan Chiras (New Society Publishers, 2009).
- *Renewable Energy Systems: The Earthscan Expert Guide to Renewable Energy Technologies for Home and Business*, by Dilwyn Jenkins (Routledge, 2012).
- *Solar Electricity Handbook: A Simple Practical Guide to Solar Energy*, by Michael Boxwell (Greenstream Publishing, 2012).
- *Solar Technology: The Earthscan Expert Guide to Using Solar Energy for Heating, Cooling and Electricity*, by David Thorpe (Routledge, 2011).
- *Grid-Connected Solar Electric Systems: The Earthscan Expert Handbook for Planning, Design and Installation*, by Geoff Stapleton and Susan Neill (Routledge, 2012).

There are also resources for those interested in learning more about larger commercial and utility-scale solar projects:

- *Solar Photovoltaic Projects in the Mainstream Power Market*, by Philip Wolfe (Routledge, 2012).
- *Large-Scale Solar Power Systems: Construction and Economics*, by Peter Kevorkian (Cambridge University Press, 2012).

These books provide information on the technology, design, and installation of solar projects. The volume of information is likely to be more detailed than you will find necessary in order to make informed investment decisions for your commercial properties, but you can nonetheless find a great deal of information written about nearly all aspects of the solar industry.

The intersection of the commercial real estate industry and the solar industry is a dynamic place to be. It opens the door to new revenue-generating opportunities in ways that few real estate professionals would have imagined even a few years ago. Armed with the information in this handbook you can continue your pursuit of solar projects that benefit your customers, your bottom line, and the environment.

If you have feedback about the content of the handbook, please email it to: solarhandbook@gmail.com

Index